Pest Control with Nature's Chemicals

Pest Control
with Nature's Chemicals

Allelochemics and Pheromones
in Gardening and Agriculture

by Elroy L. Rice

UNIVERSITY OF OKLAHOMA PRESS : NORMAN

By Elroy L. Rice

Allelopathy (New York, 1974)

*Pest Control with Nature's Chemicals: Allelochemics and Phero-
mones in Gardening and Agriculture* (Norman, 1983)

Library of Congress Cataloging in Publication Data

Rice, Elroy Leon, 1917-
    Pest control with nature's chemicals.

    Includes bibliographical references and index.
    1. Pest control—Biological control. 2. Allelopathy. 3. Insect-plant
relationships. 4. Insect baits and repellents. 5. Pheromones. I. Title. II.
Title: Allelochemics and pheromones in gardening and agriculture.
SB975.R53   1983   632'.96   83-47838

Copyright © 1983 by the University of Oklahoma Press, Norman, Publish-
ing Division of the University. Manufactured in the U.S.A. First edition.

To my students, whose enthusiasm, drive, and accomplishments have made my work a pleasant experience

Contents

Illustrations

Preface

PLANTS and animals produce organic compounds that affect the growth, health, behavior, and population biology of other plants and animals in ways other than nutrition. This book is written for scientists and general readers who desire to learn something about these important chemical interactions. Most people are familiar with the antibiotics which are produced by microorganisms, such as actinomycetes, bacteria, and fungi, and which affect the growth and survival of other microorganisms. Many are not aware that plants also produce chemicals that perform various functions, such as attracting or repelling predatory animals and stimulating or deterring feeding and egg laying. Moreover, animals produce chemicals that repel or attract other animals, causing mating, for example, or marking territories and trails.

These chemical interactions between organisms play important roles in natural ecosystems and are influential in forestry, gardening, and agriculture. The focus in this book is on the importance of chemical interactions in gardening and agriculture, but many of the examples given are relevant

to other disciplines as well. Where important chemical interactions have not been demonstrated using garden or crop
plants, examples are drawn from wild species. The use of
chemical interactions in the biological control of pests is
particularly emphasized.

It is interesting that even the ancient Roman and Greek
writers, before the time of Christ, mentioned many chemical
interactions, particularly between different plant species but
also between plants and animals. Their chief interest, of
course, was the medicinal uses of plants. Most scientific
research concerning biochemical interactions between organisms has occurred since World War II, and research has
been particularly active since 1970. There have been so many
publications that several large volumes would be required
for even a cursory review of the field. Therefore this is just
a general overview of this important subject, with selected
examples. The reader who wishes more details should study
the Notes at the end of the book, which give many citations
of the pertinent literature.

The term *plant* is used very broadly here to include fungi,
algae, and bacteria in addition to those organisms, such as
mosses, ferns, and flowering plants, that are most often considered as plants. The term *animal* is also used broadly in
the zoological sense to include all the different groups, from
the simple, one-celled protozoans to the complex mammals.
The emphasis, however, is on a few groups of animals that
are very important in agriculture, particularly the insects and
nematodes.

Common names are used throughout the book for both
plants and animals, but the scientific names are given in
parentheses the first time that the common name is used.
Scientific names have the advantage that they are the same
all over the world, while common names vary from place to
place. The scientific name of a species consists of two words
like the name of a person. The last name is listed first, how-

ever, because the first name denotes the genus, and the second the species. The scientific names are latinized, because at the time of Linnaeus, who initiated the system of naming plants and animals, Latin was the international language of scholars.

It is my sincere hope that all who are interested in gardening and agriculture, or just interested in living things in general, will find this account interesting, educational, and useful.

I am grateful to the library personnel in the Rare Books Collections in the University of Illinois, Urbana, and the National Agricultural Library, Beltsville, Maryland; and to the personnel in the History of Science Collection in the University of Oklahoma, Norman, for their help in my search for historical literature on chemical interactions between organisms. I am thankful also to Beverly Richey and Susan Nelson for typing the manuscript and to my wife, Esther M. Rice, for helping with the literature search and proofreading.

ELROY L. RICE

Norman, Oklahoma

Pest Control with Nature's Chemicals

CHAPTER 1

Historical Observations of Plant-Plant Chemical Interactions

THEOPHRASTUS in his *Enquiry into Plants,* written some-time before 285 B.C., pointed out that chick-pea *(Cicer arietinum)* does not reinvigorate the ground as other legumes do but "exhausts" it instead. Moreover, "it destroys weeds, and above all and soonest caltrop" *(Tribulus terrestris).*[1]

Pliny, in book 17 of his *Natural History,* written in the first century A.D., observed that chick-pea, barley *(Hordeum vulgare),* fenugreek *(Trigonella foenum-graecum),* and bitter vetch *(Vicia ervilia)* all "scorch up" cornland. He pointed out also that stone fruit, especially olives, should not be planted in cornland and he quoted Virgil as saying that cornland is also "scorched" by flax *(Linum usitatissimum),* oats *(Avena sativa),* and poppies *(Papaver* spp.).[2]

Pliny stated, also in book 17, that the "shade" of a wal-nut tree (apparently *Juglans regia*) is "heavy, and even causes headache in man and injury to anything planted in its vicinity" and that of a pine tree also kills grass. On the other hand, he pointed out that "the shade of the alder is dense but per-mits the growth of plants." He concluded that "each kind of plant shade is either a nurse or else a step-mother—at all events for the shadow of a walnut tree or a stone pine or a spruce or a silver fir to touch any plant whatever is undoubtedly poison." His subsequent discussions indicate that he used the word *shade* in a very broad sense to mean not only the partial exclusion of light but also the effects of plants on the nutrition of neighboring organisms and the chemicals that escape from plants into their environment. His statement that the shade of the walnut even causes headaches in man was reiterated in most of the old medical herbals, which claim that walnut trees cause headaches, dizziness, and blurred vision.

The following statement by Pliny is further evidence that his concept of shade included any volatile chemicals escaping from the plant:

> The nature of some plants though not actually deadly is injurious owing to its blend of scents or of juice—for instance the radish and laurel [*Laurus nobilis*] are harmful to the vine [*Vitis*]; for the vine can be inferred to possess a sense of smell, and to be affected by odors in a marvellous degree, and consequently when the evil-smelling plant is near it to turn away and withdraw, and to avoid an unfriendly tang. This supplied Androcydes with an antidote against intoxication, for which he recommended chewing a radish. The vine also abhors cabbage and all sorts of garden vegetables, as well as hazel [probably *Corylus avellana*], and these unless a long way off make it ailing and sickly; indeed nitre and alum and warm sea water and the pods of beans or bitter-vetch are to a vine the direst poisons.

Pliny also claimed that trees are killed by the legume cytisus *(Cytisus laburnum)* and by a plant that the Greeks called halimon (perhaps *Atriplex halimus)*.

In a discussion of how to get rid of undesirable plants in book 18, Pliny advised that the best way to kill bracken fern *(Pteridium aquilinum)* is to knock off the stalk with a stick when the stalk is budding, "as the juice trickling down out of the fern itself kills the roots." He observed also that those ferns do not survive if they are "cut with a reed [*Acorus calamus*] or plowed up with a reed placed on the plowshare." Moreover, he recommended that reeds be plowed up with bracken placed on the plowshare in order to eliminate the reeds. Another weedy pest, discussed by Pliny in book 19, was esparto grass *(Stipa tenacissima),* which he said is "a curse of the land, and nothing else can be grown or can spring up with it."

In another discussion in book 19 of chemical interactions between cultivated plants, Pliny emphasized again the "great antipathy between radishes and vines, which shrink away from radishes planted near them." He stated also that "rue [*Ruta*] is so friendly with the fig that it grows better under this tree than anywhere else." Moreover, he wrote, in book 20, that rock cabbage *(Brassica)* "is strongly antipathetic to wine, so the vine tries very hard to avoid it, or, if it cannot do so, dies." In book 18 he stated that "the first of all forms of disease" in wheat is wild oat grass *(Bromus).*

Like Pliny the ancient physician Dioscorides wrote his *Greek Herbal* in the first century A.D. He was primarily concerned with the medical uses of plants, though he made a few comments about plant-animal chemical interactions, which will be discussed in the next chapter.[3] The *De Materia Medica,* included in the *Greek Herbal,* served as the standard reference on the medical uses of plants for 1,500 years and was the basis of many, if not most, of the medical herbals written in various parts of the world after the beginning of the sixteenth century A.D. Like most of the ancient Greek and Roman writers on plants, Pliny was concerned chiefly with their uses in medicine, though he had more to say about other aspects of plant life than most of the other writers.

More than 1,400 years elapsed before other herbals appeared. In 1597, J. Gerarde published *The Herball or Generall Historie of Plantes.* In it he stated, "It is reported that the rawe Colewoort [cabbage] being eaten before meate, doth preserve a man from drunkennesse, the reason is yeelded, for that there is a naturall enmitie betweene it and the vine: which if such as it growe neere unto it, foorthwith the vine perisheth and withereth away."[4] This observation is similar to Pliny's report of Androcydes's antidote against intoxication, though, of course, radish was recommended by Androcydes rather than cabbage.

Nicholas Culpeper, in a discussion of basil *(Ocimum)* in his *English Physitian and Complete Herball,* published in 1633, complained, "Something is the matter; this herb and rue will never grow together, no, nor near one another." Like Gerarde, Culpeper observed that there is such an antipathy between grape vines and cabbage that they will not grow together. Moreover, he said, sage *(Salvia officinalis)* likes to grow not, as might be supposed, with onions but with marjoram *(Origanum majorana).*[5]

Sir Thomas Browne, in "The Garden of Cyrus," published in 1658, made the following statement concerning plant transpiration: "And as they send forth much, so may they receive somewhat in: For beside the common way and road of reception by the root, there may be a refection and imbibition from without; For gentle showrs refresh plants, though they enter not their roots; And the good and bad effluviums of vegetables, promote or debilitate each other."[6] Interestingly, this statement is generally acceptable in terms of our present understanding of plant physiology and plant-plant chemical interactions.

A Japanese writer, Banzan Kumazawa, in a document apparently written during the seventeenth century, reported that rain or dew washing the leaves of red pine *(Pinus densiflora)* is harmful to crops growing under the pine.[7] In 1832 the Swiss botanist Augustin Pyrame de Candolle, in his *Physiologie végétale,* suggested the possibility that some plants may excrete substances from their roots that are injurious to other plants.[8] He observed that thistle *(Cirsium)* in fields injures oats, while *Euphorbia* and *Scabiosa* injure flax, and rye *(Lolium)* injures wheat. He described the experiments of M. Macaire, who had found that beans will die in water that contains material exuded from the roots of other plants of the same species, whereas wheat flourishes in water charged with exudations from legumes. De Candolle suggested that such root excretions could explain the exhaustion of soil by certain plants and consequently the need for crop rotation.

In 1841 an "Editor's Note" in the British publication *Gardeners Chronicle* stated that *Rhododendron arboreum* does well under the drip of trees in a light, gravelly soil, but not in clay.[9] Like the observation of Kumazawa, this statement is remarkable in the light of our present knowledge that plant-produced toxins often concentrate in fine-textured soils to a toxic level, but often leach below the depth of rooting in coarse-textured soil. A home correspondent, listed only as T. T., described some interesting observations in an 1842 issue of *Gardeners Chronicle.* He (or she) pointed out that none of the common evergreen shrubs, such as laurels, box, rhododendrons, yews, and hollies, flourish in the neighborhood of beech trees *(Fagus),* though they have plenty of headroom and the grass is destroyed beneath those trees. The correspondent then added that it would be interesting to know whether this is so because of some "noxious quality communicated to the rain-water which drips from the beech foliage," or because beech-tree roots extract from the soil more of the materials required for growth than do the roots of larches, firs, oaks, sycamores, elms, ashes, and other hardwoods, near to which the same shrubs grow luxuriantly.[10]

A correspondent identified as "a Montrose Inquirer" had some interesting comments and questions about fairy rings in an 1845 issue of the *Chronicle:*

> In Links of Montrose there are many fairy rings of fungi and in all, I have found grass, clover, and other plants, especially

in the center of the ring, in a sickly or dying condition. The grass touching the decaying fungi was black, as if charred. There are also a great number of fairy rings upon all of which the grass has the same sickly or dying appearance; and upon those parts where the fungi have grown most numerous and luxuriant, every plant has died and left the ground quite bare. From these facts, I think it is evident that the decayed fungi cause this disease and death. Now if decayed fungi can destroy these plants, may they not also destroy the seeds [spores] of the fungi upon which they fall? Apply this principle to the small circular masses. The great number of fungi decaying in the center will destroy the seeds of the fungi upon which they fall, while those seeds falling from the fungi on the outer margin of the circle will fall where no fungi have decayed; they will be uninjured, and thus a ring will be formed.[11]

It is interesting that J. T. Way, suggested a similar explanation of fairy rings in 1847.[12]

In an 1845 *Gardeners Chronicle,* a home correspondent, named Beobachter, observed that plants on a heath form a peculiar hard-block stratum a few inches below the soil surface.[13] The stratum is equally impervious to water and to the roots of trees. Unless the stratum is broken, it is useless to plant trees, because most will die, and those that survive will not do well. The stratum is so compact and heavy that, when first dug up, it might pass for ironstone. It seems to contain a large proportion of carbonaceous matter, and, when passed through a red heat, it is converted into an incoherent red sand. Beobachter remarked that the existence of this substratum beneath the heath favors the supposition that plant roots produce excrements.

A letter by "T.A.," concerned with possible reasons for crop rotation, also appeared in the *Gardeners Chronicle* in 1845:

Let the land be supplied with all the chemical elements of vegetation in abundance, if the same crop is sown two or three years in succession, it will be found deficient; also when crops nearly allied succeed each other.—Although the cause has not been satisfactorily accounted for on chemical principles, it would appear that the excretions of the roots of one culmiferous [hollow-stemmed] plant was injurious to those of another of the same family, or that the one abstracts some peculiar principle from the soil essential to its vigorous growth. —In land that was sick of clover, instead of sowing it everytime of fallow, I have missed it once in a course, so that the interval between the crop was 7 or 8 years; this plan as far as I have observed, is attended by a complete restoration of the crop, and is the best to adopt in the present state of knowledge.[14]

An 1847 editorial in the same journal noted that "cloversickness" was a serious problem in some soils and that various explanations had been suggested. It concluded, however, that the causes were still enveloped in mystery.[15] As early as 1804, in the *Farmers Calendar,* published in London, A. Young had observed that clover is extremely apt to fail in fields where it has been constantly cultivated, because the soil becomes "sick of clover."[16]

In 1881 two Americans, J. S. Stickney and P. R. Hoy, had observed, like Pliny, that vegetation under the black walnut tree *(Juglans nigra)* is very sparse compared with the growth under other commonly planted shade trees.[17] They pointed out that no crop will grow under or very near the black walnut. Stickney questioned whether this is so because of the water dripping from the tree or because the tree, being a gross feeder, exhausts the soil. Hoy meanwhile had concluded that the main reason for the absence of vegetation under the walnut is the poisonous nature of the drip.

A letter to the editor of the *Gardeners Chronicle* in 1894 pointed out that paying crops cannot be obtained if cucumbers are grown in the same greenhouse for more than three years without other crops intervening. Discussion was invited concerning possible reasons for this, and J. J. Willis, an editor, gave the following explanation:

It seems to be the usual practice after the growth of a cucumber crop to remove all the surface-soil taking it out to a depth of 12 to 15 inches, and to convey into the cucumber-house entirely fresh soil. Hence, whatever may be the cause of failure, it cannot be attributed to the surface soil. Plants set in newly-imported soil have been known to make good healthy growth up to a certain point, and then to show signs of decay, and finally yield a meagre crop of fruit. It appears to me, therefore, that we must look for the primary cause of failure to the subsoil, and it is to that, that I have directed my attention.—From observations made of the subsoil in which cucumbers had been grown, I have found that the roots penetrate to a considerable depth, and fill the subsoil with a mass of fibrous root-matter. A soil, therefore, in which several successive crops of cucumbers have been grown naturally becomes charged with much decaying vegetable matter and the supposition is, that the primary cause of failure may be due to excreted substances given off by the roots during growth, which accumulate in the subsoil.[18]

In another article in the *Gardeners Chronicle* in 1894, Willis discussed the problem of American farmers growing peppermint *(Mentha piperita)* in Connecticut. Willis reported that the land could generally be cropped two and sometimes three years from one plant setting. After that "the soil usually becomes so foul that the quality of oil produced would not be good enough to pay for harvesting."[19]

Near the end of the last century the Illinois Horticultural Society had a running discussion in its *Transactions* concerning the planting of orchards, vineyards, and berry patches. In 1893 a Mr. Webster stated that any crop planted near to raspberries or blackberries will injure the berries.[20] He said that he had seen farmers who sowed oats along with their strawberries and let the oats fall over to protect the berries, but in nine out of ten such trials the strawberry crop was a failure. In 1895 a Mr. Austin stated: "You must not expect to raise trees and grass off the same ground. The grass will destroy the trees."[21] No reasons for the detrimental results were suggested.

It is obvious from this brief history that farmers, gardeners, and botanists have observed and surmised chemical interactions between plants for over 2,000 years, though there were no controlled scientific experiments demonstrating such phenomena until this century. It is noteworthy that, for the majority of the species suggested to have pronounced chemical effects on their own or other species, the effects have subsequently been demonstrated in controlled experiments.[22,23]

Many of the plant species that have been widely known and used for their powerful medicinal effects have pronounced chemical effects on other plants as well. Similarly, the environmental factors that cause plants to be more potent in medical use also cause the plants to have stronger chemical effects on other plant species. Dioscorides observed in his *Greek Herbal:*

> We ought to gather herbs when the weather is clear, for there is a great difference whether it be dry or rainy when the gathering is made. The place also makes a difference: whether the localities be mountainous and high, whether they lie open to wind, whether they be cold and dry; upon this the stronger forces of drugs depend. Medicinal plants found growing on plains in plashy and shady localities, where the wind cannot blow through, are for the most part the weaker; especially those that are not gathered at the right season, or else are decayed through weakness.

We know now that stress conditions make the concentrations of toxins greater in plants also. This includes such stress factors as drought, the increased ultraviolet radiation that occurs at higher elevations and during clear weather, increased light intensity, significant deficiencies of required minerals, low temperatures, and so on. Those stress conditions coincide closely with the conditions listed by Dioscorides in the first century A.D.

It appears desirable at this point to describe a few of the experiments on plant-plant chemical interactions in the twentieth century. Starting in 1907, O. Schreiner and his associates published a series of papers in which they presented evidence that single-cropping of certain plants exhausts the soil by the addition of growth inhibitors to it. Schreiner and H. S. Reed demonstrated clearly in 1907 that the roots of seedlings of wheat, oats, and certain other crop plants exude materials into the growing medium that elicit chemotropic responses by the roots of seedlings of the same species.[24] In 1908, Schreiner and Reed developed a technique that is still used to determine the possible effects on plant growth of compounds obtained from the soil or from plants.[25] They were able to show with this technique that many compounds previously identified from various plants inhibited the growth and transpiration of wheat seedlings. In 1909, Schreiner and M. X. Sullivan extracted an unidentified substance from soil that had been fatigued by cowpeas *(Vigna sinensis)* and demonstrated that it strongly inhibited cowpea growth.[26] After the inhibitor had been extracted, the soil was no longer inhibitory to cowpeas.

Between 1917 and 1919, S. V. Pickering demonstrated that the leachate from trays containing certain species of grasses inhibited the growth of apple seedlings.[27,28] The design of the experiment was such that mineral deficiencies, root interaction, shading, water deficiency, and oxygen exclusion were eliminated as possible causes of growth inhibition. Pickering's results recall Austin's statement in 1895 that grass destroys fruit trees.

In 1925, A. B. Massey did a careful study of the inhibitory effects of black walnut on alfalfa and tomato plants.[29] The test plants of both species wilted and died whenever their roots came in close contact with the walnut roots. There was no specific relationship between the region of greatest concentration of walnut roots and the wilting of the tomatoes, as one would expect if the trouble were caused by lowering of soil moisture. Apparently there was little or no poisoning of the soil, since the roots of the affected plants were in close contact with those of the walnut. When several pieces of bark from walnut roots were placed in a water culture of tomato plants, the plants wilted, and their roots browned within forty-eight hours. The addition of bark from walnut roots to soil in which tomato plants were growing caused the plants to grow poorly. In 1928, R. F. Davis extracted and purified the toxic substance from the hulls and roots of walnut and found it to be identical to juglone, 5-hydroxy-α-naphthoquinone.[30] This compound is a powerful toxin when injected into the stems of tomato and alfalfa plants. Thus the observations of Pliny in the first century and Stickney and Hoy in 1881 were confirmed in the twentieth century.

In 1932, O. H. Elmer found that the ripe fruits of four varieties of apples and one variety of pears produced volatile substances that inhibited the normal sprout development of potatoes.[31] H. Molisch carried out numerous experiments with apple fruits which confirmed and extended the work of Elmer.[32] He coined the term *allelopathy* to refer to biochemical interactions between all types of plants, including microorganisms. His discussion indicated that he meant the term to cover both the detrimental and the beneficial reciprocal biochemical interactions.

In 1940, H. R. Bode reported that the foliar excretions of wormwood *(Artemisia absinthium)* inhibited the growth of seedlings of several plant species within one meter of the wormwood plants.[33] According to Bode, the glandular hairs on wormwood leaves excrete volatile oils and the inhibitor absinthin. These substances appear as numerous droplets on the surface of the hairs, and, when it rains, the droplets are washed away and spread onto neighboring plants. In 1943,

13

G. L. Funke confirmed and extended Bode's results.[34] He determined the effects of a wormwood hedge on a large number of test species planted near it and found that all were affected. No effect was observed on the same species when they were planted near a hedge of a different plant, *Atriplex hortensis.* He found also that, if fresh or pulverized wormwood leaves were dug into the soil, they retarded the germination of pea *(Pisum sativum)* seeds and permanently lowered the percent of bean seeds that germinated. The growth of bean plants in the soil was also permanently retarded by these leaves. In later experiments the seed germination and seedling growth of numerous other species were found to be severely inhibited in soil in which wormwood leaves had been incorporated.

H. M. Benedict observed that pastures of smooth brome *(Bromus inermis)* become drastically thinned after two or three years of growth.[35] He found that, when oven-dried roots of the species were placed in soil with seeds of the species, there was a significant reduction in the subsequent dry weight of the seedlings as a result. Similar results were obtained when he added a water leachate from an old culture of smooth brome to seedlings of this species. He concluded therefore that the thinning of the pastures was caused by the autotoxic effect of substances produced by the smooth-brome roots.

There are many hundreds of research projects in allelopathy that are beyond the scope of this book. Some of the more pertinent modern projects concerned with agriculture will be discussed in Chapter 3. Fortunately, the pace of research in this field is accelerating rapidly, particularly in relation to agriculture and forestry.

CHAPTER 2

Historical Observations of Plant-Animal Chemical Interactions

ABOUT 300 B.C., Theophrastus, in book 9 of his *Enquiry into Plants,* stated, "Those roots which contain any sweetness become wormeaten in course of time, but those that are pungent are not so affected, though their virtues diminish as they become flabby and waste away."[1] He pointed out also that wolfsbane, or scorpion plant *(Aconitum anthora),* kills scorpions if it is shredded over them.

In book 20 of his *Natural History,* Pliny advised that a mallow leaf *(Malva)* "placed on a scorpion paralizes it," that "snakes are kept away by the sawdust of cedrus [*Juniperus excelsa*], and that to rub the body with the crushed berries mixed with oil has the same result." In book 12 he reported that the fruits and leaves of citron, or Assyrian apple (*Malus Assyria* in the Latin text), have an exceptionally strong scent that "penetrates garments stored with them and keeps off injurious insects."[2] He observed that the tree bears fruit at all seasons, some of the apples falling while others are ripening and still others are just forming.

In his book 17, Pliny agreed with Theophrastus: "The trees most likely to be worm-eaten are pears, apples, and figs;

those that have a bitter taste and a scent are less liable." He reported that lupine seeds can be left lying on the ground with impunity, as they are protected from animals by their bitter flavor as long as a rain does not wash away the bitter substance. In Pliny's time caterpillars were a problem for all cultivated crops. In book 18 he recorded that they even attacked chick-pea "when rain makes it taste sweeter by washing away its saltness." He stated that the seeds of cereals "are not liable to maggots if mixed with crushed cypress leaves" and that "the most effective thing for killing ants is the heliotrope plant [*Heliotropium*]." Moreover, he claimed that caterpillars were repelled if the forage crops fitch *(Vicia sativa)* and rape *(Brassica napus)* were sown together, and if chick-pea was sown with cabbage. He claimed that, "if cabbage seed is soaked in the juice of houseleek [*Sempervivum tectorum*] before being sown, the cabbage plants will be immune from all kinds of insects."

J. Gerarde, in his *Herball or Generall Historie of Plantes,* published in 1597, reported that the roots of avens *(Geum),* "taken up in Autumne and dried, do keepe garments from being eaten with Mothes, and make them to have an excellent good odour: and serve for all the physicall purposes that Cinkefoiles do."[3] Nicholas Culpeper, in his *English Physitian and Complete Herball,* published in 1633, recommended wormwood for the same purpose: "Wormwood being laid among cloathes will make a moth scorn to meddle with clothes as much as a lion scorns to meddle with a mouse or an eagle with a fly." He stated also that hot arsmart (probably *Polygonum* sp.), "if strewed in a chamber will soon kill all the fleas." Moreover, if the herb or juice of mild arsmart is applied to horses, it will drive away flies, according to Culpeper. He also advised his readers, "Mix a little wormwood with your ink and neither rats nor mice will touch the paper written with it."[4]

W. Coles, in his book *Adam in Eden: or Natures Para-*

16

dise, published in 1657, stated that smoke from burning stalks of loosestrife *(Lythrum salicaria)* "driveth away serpents, or any other venemous creature, as Pliny faith [*sic*] and the people in the Fenny Countreyes can testifie that it driveth away the Flyes and Gnats, that would otherwise molest them in the night season."[5] The indication here is that this statement was made by Pliny in his *Natural History,* but I did not note it in my reading of that massive work. Coles described other natural pesticides as well. Water in which rue *(Ruta)* has been soaked, if scattered about the house, will drive away fleas and kill them. The galls of a sumac will keep moths from garments and woolen clothes, "giving unto them a good scent, and therefore it is much used to be laid in Wardrobes, Chests, Presses, and the like." The juice from marigold flowers *(Tagetes)* "dropped into the Ears, killeth Worms," and the oil from hyssop *(Hyssopus)* "killeth lice."

In *The English Physician Enlarged,* published in 1681, Culpeper added several examples of plant-animal biochemical interactions that he did not mention in his earlier edition.[6] He stated that, if common alder (*Alnus* sp.) leaves are gathered while the morning dew is on them "and brought into a chamber troubled with fleas, [the leaves] will gather them thereunto, which being suddenly cast out, will rid the chamber of those troublesome bedfellows." He suggested that smoke from a burning fern would drive away "serpents, gnats, and other noisome creatures," and, like Coles, he advocated smoke from burning loosestrife (*Lythrum* sp.) to drive away flies and gnats. He recommended tobacco juice to kill lice on children's heads, a very early reference to the use of tobacco as an insecticide.

E. G. Beinhart, writing almost three centuries later, in 1950, gave a history of the production and use of nicotine. Tobacco and its extracts were used for the control of insects long before it was known that nicotine was the toxic agent.[7] According to Beinhart, continental and English gardeners early recognized the properties of the tobacco that was imported from the American colonies. In a letter dated January 20, 1734, Peter Collison of London suggested to an American correspondent, John Bartram (a Philadelphia botanist), the use of tobacco leaves to protect the letters and packages containing seeds and plants that were being shipped to Collison. In 1746, Collison advised Bartram to use a water extract of tobacco for the control of the plum curculio on nectarine trees. Tobacco dusts and extracts were recommended for the control of plant lice in France in 1763, and in America, in 1832, Thomas Fessenden included tobacco in a list of insect repellents and insecticides. In 1884 tobacco was described as one of the three most valuable insecticides in general use, the other two being white hellebore *(Veratrum album)* and soap.

Jared Eliot, in his *Essays upon Field-Husbandry in New England,* published in 1760, described a technique to prevent birds from eating corn grains after they are planted: "Boil roots of swamp hellebore (sometimes called skunk cabbage, tickle weed, bear root [*Symplocarpus foetidus*]) for two hours in water which covers them an inch deep. Soak corn grains in the warm liquor for 20 hours and plant."[8]

Samuel Deane advised, in *The New-England Farmer; or Georgical Dictionary* (1790), "To destroy lice on cabbages, they should be washed with strong brine, or seawater, or smokes should be made among them with straw, sulphur, tobacco, etc."[9] He reported that he had kept caterpillars out of his orchard by hanging rockweed (*Fucus,* a seaweed) in the crotches of the trees in the spring of the year. A friend of Deane's kept butterflies away from his cabbages by whipping the plants gently with elder branches just as the butterflies appeared. Although the butterflies hovered over the cabbages, they were never observed to touch them. The friend also whipped the limbs of a plum tree as high as he could reach, and that part of the tree "remained green and flourishing, whereas all above shriveled up and was full of worms." He also prevented the yellows in wheat—a disease that is caused by an insect—by brushing the wheat with elder and preserved a bed of young cauliflowers in the same manner. According to Deane, the friend preferred the dwarf elder for repelling insects, because it "emits a more offensive effluvium." Deane reported that, if water in which walnut leaves have been steeped for two or three weeks is sprinkled on gardens, it will "subdue the worms." In 1830 an editorial comment in the *New-York Farmer and American Garden Magazine* stated: "Elder leaves put around the roots of peach trees, is recommended as a perfect antidote for the injury arising from worms."[10]

C. S. Rafinesque, in his *Medical Flora; or Manual of Medical Botany of the United States of North America,* published in 1830, reported that the leaves and wood of all junipers *(Juniperus)* contain a resin, which he called sandarac, that renders the wood very durable and obnoxious to insects.[11]

A brief article appearing in 1830 in the Miscellaneous Sections of *New-York Farmer and American Garden Magazine,* reported that, where acorn squash seeds are planted liberally with cucumber and melon, insects prefer feeding on the squash seeds and neglect the others.[12] The same magazine carried the following editorial comment in 1834: "Snails, worms and the grubs or larvae of insects, as well as moles and other vermin, may be driven away by placing preparations of garlic in or near their haunts."[13]

P. Mackenzie wrote the following brief letter to the editor of the British *Gardeners Chronicle* in 1845: "This season I observed a number of bees *(Bombus terrestris* and *Bombus muscorum)* in a sluggish state about a plant of the Bear's-foot *(Helleborus foetidus);* they were unable to fly away to any great distance when they came from the flowers; some of them had enough to do to walk on the ground and others tried to use their wings and rose 2 or 3 feet into the air and tumbled back again into the plant."[14] Obviously, there was a strong chemical interaction in this situation, and the editor commented that bear's-foot produces a potent narcotic substance. Today alkaloids derived from hellebores are used as cardiac and respiratory depressants and insecticides.

An editor's note in an 1895 issue of the *Gardeners Chronicle* stated that the bark of the tree of heaven *(Ailanthus altissima)* "is intensely bitter, and is used in dysentery and as a vermifuge. It would in all probability be as good an insecticide as quassia [*Quassia amara*]."[15] This suggestion proved to be correct; the toxic chemicals in both plants have since been found to be closely related.[16]

One of the very important plant-animal chemical interactions in agriculture concerns insect pollination of plants. The development of our present understanding of this phenomenon took about 2,000 years, and it seems desirable now to spot some of the highlights, though the chronological order of our other historical accounts is interrupted. In the fourth century B.C., Aristotle recorded, in his *Historia Animalium:* "On each expedition the bee does not fly from a flower of one kind to a flower of another, but flies from one violet, say, to another violet and never meddles with another flower until it has got back to the hive."[17] This was an important observation even though the ancients were not aware of pollen and its importance in plant reproduction. An eighteenth-century Irishman, named Arthur Dobbs, was the first person to observe and clearly describe insect pollination and its importance.[18] Having worked with bees, he corroborated Aristotle's observation that they visit only one kind of flower on each trip from the hive. Later, in 1821, François Huber demonstrated that bees are attracted to flowers by their scent. This was an important advance in our knowledge of chemical interactions between plants and animal.[19] In his discussion of the pollination of an orchid in 1862, Charles Darwin concluded: "As the flowers [of *Orchis pyramidalis*] are visited both by day and night-flying Lepidoptera, I do not think it is fanciful to believe that the bright-purple tint . . . attracts the day-fliers, and the strong foxy odour the night-fliers."[20] Sixty years later, in 1902, Darwin stated this more strongly: "The odours emitted by flowers attract insects. . . . So great is the economy of nature, that most flowers which are fertilised by . . . nocturnal insects emit their odour chiefly or exclusively in the evening."[21]

James Rodway, in his fascinating book *In the Guiana Forest,* published in 1895, reported that apparently there were no wind-pollinated trees in the forest and proceeded to discuss the methods of pollination.

Without living helpmates many a tree would become extinct, therefore every effort is put forth to attract and induce winged creatures to render this assistance. The principal means to this end are colours and perfumes, the former for diurnal and the latter for nocturnal insects. Perhaps the most interesting point in connection with these perfumes is that they are distilled at certain times, and then only for short periods. Sometimes the flower opens, carries on its work for an hour or two, and then closes, either altogether, or in a few cases, to repeat the process at the same hour next day. It might be thought at first that these alternations are erratic, but close observation shows that they are nearly true to the minute, and if carefully timed would almost certainly be found to coincide with the period when the fertilising agent is on the wing. For this is the simple explanation; the flower can only be fertilised by a particular kind of insect, and all its efforts are put forth when that insect is likely to be hovering around. Without the flowers the bee could not exist, and without the bee no seed would be produced.[22]

Huber had reported in 1821 that freshly excised bee stingers, or their odor, when placed near the entrance to the hive, elicited aggressive attacks by the worker bees. This is an example of an early recorded observation of an animal-animal chemical interaction. As is true of plant-plant chemical interactions, detailed research on plant-animal and animal-animal chemical interactions only began about 1950.

In 1932, a German scientist, A. Bethe, introduced the term *ectohormone* for substances, such as sex attractants, that are external, rather than internal, secretions of animals and that act as chemical messengers between individuals rather than between different parts of the same individual.[23] This is a self-contradictory term, and in 1959, P. Karlson and A. Butenandt suggested the term *pheromone* for such chemicals.[24] They restricted the use of the term to substances that are secreted to the outside by an individual and then received by a second individual of the same species, in which they release a specific reaction. As restricted, the term could not be applied to the chemicals produced by plants that affect animals. R. H. Whittaker suggested the term *allelochemics* for chemicals by which organisms of one species affect the growth, health, behavior, or population biology of organisms of another species, excluding substances used only as foods by the second species.[25] This broad term obviously covers allelopathy also, except for chemical interactions between individuals of the same plant species.

CHAPTER 3

Allelopathy in Agriculture

THE allelopathic effects of crop plants on other crop plants have concerned farmers through the ages, as Chapter 1 describes. In our century most studies of allelopathy relating to agriculture have investigated the effects of decomposing crop residues, probably because of the increased use of stubble-mulch farming since the dust-bowl days of the 1930s. Such mulches help control erosion, but also decrease crop yields in many instances.

Decomposing Crop Residues. T. M. McCalla and F. L. Duley reported in 1948 that stubble-mulch farming reduces the stand and growth of corn in Nebraska under some conditions and the effect is more pronounced in rainy years.[1] Subsequently, those researchers found that mulching the soil with wheat straw at the rate of 2 to 4 tons per acre reduced the average germination of corn from 92 to 44 percent. Additional work with various crop residues caused McCalla and Duley to conclude that the inhibitive effects of the various mulches result from a combination of toxins present in the plant material plus toxins produced by microorganisms that are stimulated to grow more luxuriantly by material in the residues.[2] In 1963, F. A. Norstadt and McCalla reported that one fungus that is stimulated in stubble-mulch plots produces a potent toxin, patulin, that markedly inhibits the growth of wheat plants.[3] J. R. Ellis and McCalla found in 1973 that this fungus constitutes 90 percent of the total fungal population in soil where stubble-mulch wheat farming occurs.[4] They discovered that a single 100-parts-per-million application of patulin to soil in which Lee spring wheat had been allowed to grow to maturity produced yield reductions similar to those that occur in stubble-mulch farming.

W. D. Guenzi and McCalla used several solvent systems and techniques to extract soil from stubble-mulched and plowed plots at Lincoln, Nebraska.[5] They found that all of the fractions from the stubble-mulched plots were inhibitory to the growth of wheat seedlings, and concentrations of several identified phenolic toxins were higher in the mulched plots than in the plowed plots.

Z. A. Patrick and L. W. Koch investigated the effects of decomposing residues of timothy, corn, rye, and tobacco on the respiration of tobacco seedlings.[6] They found that chemicals were formed during the decomposition of all four species that inhibited respiration in the seedlings. Greater inhibition resulted when decomposition occurred under saturated soil conditions. The period of time required for toxic substances to be formed was markedly affected by the stage of maturity of the plant residues added to the soil. When residues from young plants were added, toxic substances were produced relatively early in decomposition but also were broken down relatively fast. When mature plant materials were added, a longer period of decomposition was required before toxic materials were formed, but the toxicity remained high for a longer period. In all instances the toxic chemicals disappeared much more rapidly from the tobacco seedlings than from the timothy, rye, and corn. Greater amounts of toxins were produced as the soils became more acid.

When seedlings were placed in toxic extracts of decomposing plant residues, Patrick and his colleague found that the root tips (that is, the growing points) soon turned brown, and formation of root hairs was limited. It is especially significant that the roots of seedlings appear to be most sensitive to the toxins in decomposing residues. They are most likely to be in contact with the localized zones of toxin production associated with bits of plant residue. The reduction in respiratory activity in the seedlings probably retards many other processes associated with growth and development.

Research on the effects of decomposing crop residues was extended to field studies in the Salinas River valley in California,[7] where soils containing crop residues were obtained from fields that had been treated in the conventional way by growers. In some fields cover crops of barley, rye, or wheat, with or without vetch, had been disked or plowed under. The amount of green crop residue averaged 10 to 15 tons per acre. The plowing or disking was done just before the plants came into full head. In other fields broad beans, sudan grass, or remnants of a commercial crop of broccoli had been disked or plowed under. Soil samples were collected from the fields periodically and were divided into three fractions: soil and residue in the relative proportions found in the field, soil after all recognizable plant residues were removed, and plant residue free of soil. Water extracts were made of each fraction and tested against the seed germination and seedling growth of lettuce. In some cases, broccoli, white beans, and tobacco were tested also.

The extracts of soil and residue in the proportions found in the field and the extracts of soil after the residue was removed showed low toxicity, whereas the residue itself had

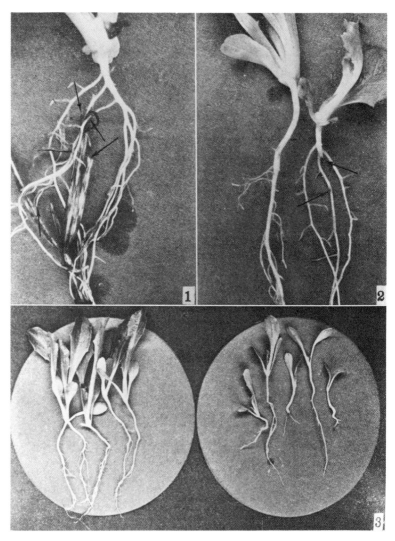

Fig. 1. 1, lettuce seedling with fragments of plant residue intermingled with roots. Arrows indicate necrotic lesions on the roots in contact with residues. Many such cases were examined in the same field. Moreover, no pathogens were found in the necrotic lesions. *2,* other lettuce seedlings from the same field. *On the right,* the seedling is from an area relatively high in decomposing barley residues; the arrows indicate necrotic lesions on the roots and a characteristic necrosis of the apical

an appreciably depressive effect on the root growth of lettuce. Soil immediately adjacent to residue and extracts of soil from the same location were found to be quite inhibitory to the root growth of lettuce (fig. 1). Overall, extracts of decomposing field residues of barley, rye, broccoli, broad bean, wheat, vetch, and sudan grass were found to be toxic to lettuce seedlings.

Careful observations were made of lettuce and spinach plants growing in the fields to determine whether the stunting, uneven growth, and root injury often observed in the Salinas valley could be caused by decomposing crop residues. It was found that roots in contact with or close to fragments of decomposing crop debris often had discolored or sunken lesions where the roots came in contact with the residues. There were numerous instances of browning of the root tips and other injuries. Apparently, the toxins do not move far from the place where they are produced; the extent of root injury and the total effect on a new crop would depend on how frequently the growing root systems of the plants encounter fragments of crop residues in which toxic chemicals are present. Patrick and his colleagues concluded that one of the most important roles of the toxins produced by decomposing crop residues is the promotion of root diseases. This phenomenon will be discussed later in this chapter.

---

meristem of the taproot. Many seedlings from that part of the field were stunted. *On the left,* the seedling is from the same field but from an area relatively low in decomposing barley residues. *3,* lettuce seedlings grown with and without barley residues. *On the right,* the seedlings were grown in field soil where barley residues had been allowed to decompose for 22 days before seeding. Roots were stunted, and necrosis of the taproot apical meristem was common. *On the left,* the seedlings were grown in the same soil and treated in a similar manner, but the field did not contain decomposing barley residues. Reproduced from Z. A. Patrick et al., "Phytotoxic substances in arable soils associated with decomposition of plant residues," *Phytopathology* 53 (1963): 152-61, courtesy of American Phytopathological Society, Saint Paul, Minn.

Fig. 2. Reduction in growth of sorghum following a previous sorghum crop in a sandy soil in West Africa: *on the left,* growth of sorghum following peanuts; *on the right,* growth of sorghum following sorghum. Reproduced courtesy of Y. Dommergues.

It has been observed for some time in Senegal in West Africa that, where one crop of sorghum *(Sorghum vulgare)* follows another, the growth of the second crop decreases markedly in sandy soils but not at all in soils high in montmorillonite (fig. 2). W. Burgos-Leon observed the same results in sorghum seedlings when roots or tops of the same species were added to sandy soil in laboratory experiments.[8] No inhibition resulted, however, when the residues were added to soil high in montmorillonite. Water extracts of roots or tops inhibited growth of sorghum and rye-grass *(Lolium perenne)* seedlings.

Using sterile techniques, Leon demonstrated that inoculation with either of two fungi, *Trichoderma viride* or an unknown species of *Aspergillus,* eliminated in a short time the inhibitory effects of aqueous extracts of sorghum roots on sorghum seedling growth. The effectiveness of the *Aspergillus* species in preventing the toxicity of water-soluble materials from sorghum was observed in the laboratory. In subsequent experiments, using uninoculated, nonsterile field soil, several weeks were required to detoxify the soil after the addition of root residues of sorghum. Leon concluded that the native microflora (bacteria and fungi) in the sandy soils of Senegal are not able to detoxify the soil fast enough to prevent the inhibition of subsequent crops of sorghum in the same soil. The clay soils apparently have sufficient bacteria and fungi of the right kinds to break down the toxins present in sorghum residues. This is an important example to keep in mind, because toxins in soils are usually inactivated eventually. Whether an inhibitory effect occurs depends therefore on the relative rates at which the toxins are added and inactivated.

Rice Paddies. Soils of rice paddies in Japan and India have been found to contain concentrations of aliphatic acids high enough to inhibit the growth of the rice.[9] The anaerobic conditions that prevail in the soils of the flooded rice paddies are very favorable to the microbial synthesis of organic acids. The unharvested parts of rice plants are customarily mixed with the soil, by plowing or some other mechanical manipulation, because this is thought to be beneficial. It has been observed commonly, however, that the second rice crop of the year in a paddy is less than the first crop. Research has demonstrated that aqueous extracts of decomposing rice residues in soil inhibit the growth of rice and lettuce seedlings and that the toxicity persists for four months.[10] The toxins were identified and quantified later in rice fields, and the amounts were greater in paddies in which rice stubble was left than in paddies from which the rice stubble was removed.

Large amounts of chemical fertilizers, including nitrogen, are required to maintain good productivity in rice paddies.[11] It would obviously be important economically if a considerable portion of the nitrogen could be furnished by biological nitrogen fixation. Conceivably this could be accomplished either by rotating legume crops with rice, as described below, or by inoculating paddies with appropriate blue-green algae, if conditions conducive to nitrogen fixation could be maintained. It has been estimated that free-living blue-green algae add from 13 to 70 pounds of nitrogen per acre per year to rice paddies. Blue-green algae associated with the tiny water fern *Azolla* in rice paddies gave yields 50 to 100 percent greater than those obtained in adjoining paddies where *Azolla* was absent.[12] In 1978, C. Y. Huang found that inoculation of pots of rice plants with blue-green algae increased grain production by 34 and 41 percent, depending on the rice cultivar used.[11]

The quantities of available leachable nitrogen are lower in paddies where rice stubble is left than in paddies where the stubble is removed.[13] This suggests that the decomposing rice residues inhibit nitrogen fixation or the rate of decomposition of organic matter, or both. In southern Taiwan soybean crops or other legumes are commonly planted immediately following a crop of rice. Soybean yields are increased by several hundred pounds per acre when the rice straw is not allowed to remain in the field and decompose, but instead is burned before the planting of the soybeans. This also suggests that decomposing rice straw inhibits nitrogen fixation, and, in fact, subsequent research has demonstrated that the known phenolic toxins from decomposing rice straw inhibit the growth of nitrogen-fixing bacteria. The toxins also inhibit nodulation and hemoglobin formation in two varieties of beans—Bush Black Seeded and Four Season.[14] The hemoglobin content of the nodules correlates well with the amount of nitrogen fixed. Sterile water extracts of decomposing rice straw in soil are also very inhibitory to nitrogen-fixing bacteria (fig. 3) and to nitrogen fixation by bacteria in the root nodules of bean plants.

Most of the five known toxins from decomposing rice straw inhibit the growth of at least one nitrogen-fixing alga, *Anabaena cylindrica,* in a relatively dilute solution, 0.003 M.[15] This is a common alga in rice paddies. Three of the five toxins significantly inhibit nitrogen fixation by *Anabaena cylindrica* in a 0.003 M concentration, and a combination of the five toxins, each in a 0.003 M concentration, completely eliminates nitrogen fixation by this alga. Some of the toxins are known to occur in rice-paddy soil containing decomposing rice straw, in concentrations as high as 0.02 M. The evidence is very strong therefore that decomposing rice straw in paddies is inhibitory to nitrogen fixation by legumes and by free-living blue-green algae.

In soil that has not had recent additions of plant residue or other organic material, microbial respiration usually proceeds at a slow rate. Moreover, fungi apparently exist mostly as spores in a state of fungistasis. This microflora usually responds to the addition of plant residue by germination, increased respiration, and growth. J. D. Menzies and R. D.

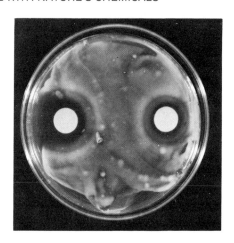

Fig. 3. Inhibition of *Rhizobium leguminosarum* by a sterile water extract of decomposing rice straw in soil. Sensitivity disks were saturated with the sterile extract and placed on petri plates inoculated with *Rhizobium.* Photograph by author.

Gilbert reported that those responses were induced by a volatile component that is found in alfalfa tops *(Medicago sativa),* corn leaves, wheat straw, bluegrass clippings *(Poa pratensis),* tea leaves *(Thea sinensis),* and tobacco leaves, even when the residue was separated from the soil by a 5-centimeter air gap.[16] There was a rapid outgrowth of hyphae from the soil surface toward the residue before any growth of fungi could be seen in the plant material. After twenty-four hours a dense network of hyphae could be detected, with many of the filaments oriented at right angles to the soil surface and reaching almost across the air gap. This development did not occur where the residue was omitted or was replaced by moist filter paper.

Vapors from distillates of water extracts of the various plant residues had similar effects on the growth of fungi, and they markedly increased the numbers of bacteria and the respiratory rate in soil. This fascinating allelopathic phenomenon may be very important in the initial colonization of residue and thus in decomposition.

So far in this chapter only the effects of decomposing crop residues have been discussed. Other examples would not add much to the general picture of allelopathic effects in agriculture. When a plant produces toxins, they may be released from its tops or roots into the environment by evaporation (if they are volatile), by exudation from living roots, by being leached from living tops by rain, or by decomposition of the plant's residue. There are many known examples of all of those methods of toxin dispersal, but most concern non-crop plants.

Volatile Excretions. V. P. Dadykin and his coworkers studied the volatile excretions of several crop plants and found acetaldehyde, propionic aldehyde, acetone, methanol, and other unidentified compounds in volatile secretions of beet, tomato, sweet potato, radish leaves, and carrot roots.[17] Propionic aldehyde was found to have the greatest activity against test species in closed systems.

Although individually the volatile compounds from the tops of soybeans, chick-peas, and beans have been reported to reduce the uptake of phosphorus (^{32}P) by corn plants, the combined volatile compounds from the tops and roots of the same three species stimulate the uptake of ^{32}P by corn.[18] In laboratory experiments the overall effect on ^{32}P uptake agreed with observed yields of mixed sowings of these plants under field conditions.

During rains a toxin is leached from the foliage of chrysanthemums *(Chrysanthemum morifolium)* that completely inhibits germination of lettuce seeds. It is also a potent inhibitor of chrysanthemum growth and development.[19] This species cannot be grown in the same soil for several years, apparently because of the accumulation of toxins.

Root Exudates. As indicated in Chapter 1, the phenomenon of clover sickness has long been known. The sickness is particularly conspicuous in red clover *(Trifolium pratense)*. In an effort to explain the causes of the autotoxicity, S. Tamura and his coworkers isolated and identified nine inhibitory iso-flavonoids or related compounds in red-clover tops.[20,21] All of the compounds were subsequently found to inhibit the seed germination and seedling growth of red clover in very dilute solutions, and they inhibited red-clover growth in soil for a lengthy period.[22] In soil the isoflavonoids are broken down to simpler compounds that are also very toxic to the clover seedlings. Thus "red clover sickness" is caused by the accumulation of toxins which result from those that exude from red-clover roots.

A soil-sickness problem also results from continuous growth of berseem *(T. alexandrinum)*. In this case chemicals exuded from the roots of berseem cause a reduction in "P-promoting" microorganisms and a marked decline in available phosphorus in the soil.[23] This sickness cannot be corrected by fertilizer applications.

Roots of two legumes, pea *(Pisum arvense)* and hairy vetch *(Vicia villosa),* exude substances that apparently stimulate both photosynthesis and the absorption of phosphorus by barley and oat plants.[24] The substances also stimulate those cereals to take in nitrogen, potassium, and calcium from nutrient solutions. In contrast, active substances exuded from the roots of the two cereals inhibit the same processes in the legumes.

Soviet scientists have done research for many years on chemical interactions of legumes and several other crop plants. They have found that some species and varieties of legumes are detrimental in mixed cultures and others are beneficial, beyond just the improvement of nitrogen nutrition by the legumes. A given legume may be beneficial in one kind of mixture and detrimental in another. Appropriate combinations have been determined for certain areas of the Soviet Union, and new varieties of legumes have been developed specifically for use in mixed culture with corn and several other forage crops. This type of research appears to hold excellent promise in agriculture.

ALLELOPATHIC EFFECTS OF WEEDS ON CROP PLANTS

Most research on the effects of weeds on crop plants purports to be confined to the competition of the weeds for minerals, water, or light. Possible biochemical interactions have received relatively little attention, although, in fact, no experiment on "competition" has been designed so that it eliminates possible allelopathic effects. Fortunately, research directly concerned with biochemical effects is increasing. There is no doubt that competition is involved, to some extent, in all the effects of weeds, but evidence is growing that allelopathic effects are very significant in at least some instances.

It has been known for many years that the yield of flax is greatly reduced when even a small percentage of flax weed *(Camelina alyssum)* is growing among the flax plants. No toxic root exudates are apparent, but the flax weed leaves are the source of a potent inhibitor.[25] Under artificial rain flax plants in close proximity to flax-weed plants produced 40 percent less dry matter than the controls, for which the same amount of water was added directly to the soil. The setup was such that there was no competition for minerals, water, or light.

Water extracts of peppergrass *(Lepidium virginicum),* evening primrose *(Oenothera biennis),* and crabgrass have been reported to be very toxic to the seed germination of crown vetch *(Coronilla varia).*[26] Peppergrass residues that remain incorporated in soil for ten weeks are also toxic to the germination of crown-vetch seeds.

The growth and yield reductions of corn infested by giant foxtail *(Setaria faberii)* have been well documented.

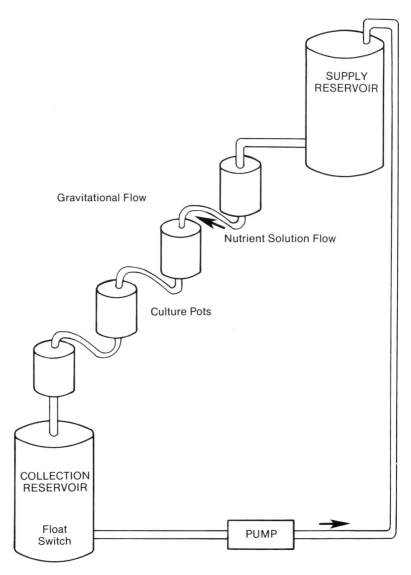

Fig. 4. Stair-step apparatus. A diagrammatic representation of one line of the experimental apparatus used to separate mechanisms of allelopathy from those of competition. Test lines contained pots of corn alternating with pots of *Setaria faberii.* Control lines contained only pots of corn. Reproduced from D. T. Bell and D. E. Koeppe, "Noncompetitive effects of giant foxtail on the growth of corn," *Agron. J.* 64 (1972): 321–25, courtesy of American Society of Agronomy, Madison, Wis.

The reductions were ascribed to competition, without any effort to find out if allelopathic effects are involved, until M. M. Schreiber and J. L. Williams demonstrated that decaying giant-foxtail roots markedly inhibited the growth of corn roots even when sufficient nitrogen was added to reduce the short-term effect of a high carbon-nitrogen ratio.[27] David Bell and D. E. Koeppe decided to investigate the effects of giant-foxtail root exudates under conditions in which competition was not involved (fig. 4).[28] Their results indicated that exudates of giant-foxtail seedlings did not affect the growth of corn plants, but exudates of mature roots, leachates of dead roots, and leachates of giant-foxtail whole-plant residue significantly reduced the height, the accumulation of dry weight, and the accumulation of fresh weight of corn plants (fig. 5). They found that, when competition was removed from the interference of giant foxtail with corn, the inhibition of corn growth dropped from 90 to 35 percent, compared with weed-free controls. They concluded that the 35-percent reduction in corn growth under these conditions could be attributed only to allelopathic compounds.

Water extracts of many weed seeds inhibit the germination of seeds of certain crop plants.[29] Moreover, incorporation of certain weed seeds in soil with crop-plant seeds inhibits the elongation of the crop seedlings.

Piru *(Schinus molle)* is a pernicious weed associated with crop plants in some parts of Mexico. A. L. Anaya and A. Gomez-Pompa demonstrated that extracts of the leaves and fruits of this weed strongly inhibit the germination and seedling growth of cucumber and wheat.[30]

Some areas of Galicia, Spain, are occupied by various species of heath (Ericaceae). When the lands are used for agriculture, the crop plants suffer serious growth problems.[31,32] Ten species, distributed in three genera, were tested for inhibitory activity against oat seedlings, and all were found to be allelopathic against oats.

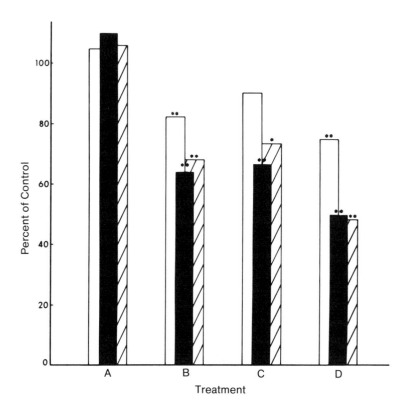

Fig. 5. Allelopathic influence of *Setaria faberii* on *Zea mays.* Height *(white bars),* fresh weight *(black bars),* and dry weight *(hatched bars)* are presented as a percentage of the control after one month's association. Treatments included corn seedlings *(A)* started together with giant foxtail seedlings, *(B)* growing with mature live giant foxtail, *(C)* growing with whole dead giant-foxtail plants, and *(D)* growing in contact with the material leached from giant-foxtail detritus, which was cut and incorporated into the sand-culture pots. *Significant difference from the control at the 0.05 level. **Significant difference from the control at the 0.01 level. Reproduced from D. T. Bell and D. E. Koeppe, "Noncompetitive effects of giant foxtail on the growth of corn," *Agron. J.* 64 (1972): 321-25, courtesy of American Society of Agronomy, Madison, Wis.

Leaves of curly dock *(Rumex crispus)* inhibit the growth of corn and grain-sorghum seedlings.[33]

In Santiago de Compostela, Spain, nut grass *(Cyperus esculentus)* interferes strongly with corn crops, reducing production and preventing the growth of other plants in the field. R. S. Tames and coworkers found that extracts of the nut-grass tubers contained several compounds that inhibited the growth of oat seedlings and the seed germination of many crop plants.[34]

Much research has been done over the years on the allelopathic effects of couch, or quack, grass *(Agropyron repens)* on crop plants. K. P. Buchholtz observed that corn plants growing in areas infested with couch grass appear to be suffering from a severe deficiency of mineral elements, particularly nitrogen and potassium.[35] Analysis of corn plants from such areas demonstrated that they were indeed low in nitrogen and potassium, compared with plants from noncouch grass areas. Yet heavy fertilization with nitrogen and potassium in couch-grass areas did not improve the yield of corn greatly, even though it was found that only a small portion of the added elements was absorbed by the couch grass. Subsequent research suggested that couch grass probably caused the nutrients in the soil to be made unavailable in some way, or it decreased the absorptive capacity of the corn roots, or both.

In 1974, J. Minar added to the evidence in the couch-grass story. He found that couch grass reduced the production of both fresh and dry matter in wheat tops.[36] The addition of fertilizer could not overcome the effect of couch grass completely, even when other possible competitive mechanisms were effectively eliminated. Subsequent research indicated the effect of couch grass is chiefly to reduce the phosphorus uptake of wheat, even when phosphorus is available in adequate amounts.

39

The growth of annual thistles in southern Tasmania is restricted to areas not colonized by Canada thistle *(Cirsium arvense)*, which is a serious weed in field crops there. According to G. M. Bendall, extracts of the roots and foliage of Canada thistle and of the decaying roots or tops inhibit seed germination in several crop plants. Decaying roots are more inhibitory than decaying tops.[37]

The common milkweed *(Asclepias syriaca)* is a major weed in the north-central and northeastern United States and Canada. In tests in Nebraska it significantly reduced the yield of grain sorghum, and James Rasmussen and Frank Einhellig found that water extracts of milkweed leaves significantly inhibited the growth of grain-sorghum seedlings.[38] Reducing the concentrations of the extracts resulted in proportional increases in yield.

The common sunflower *(Helianthus annuus)* is a serious crop weed in Mexico and several other countries, including the United States. Most of its detrimental effects have been attributed to competition, but R. E. Wilson and E. L. Rice[39] found it to be strongly allelopathic to many species of plants (fig. 6).

D. Gajić and her colleagues in Yugoslavia found that grain yields are increased appreciably over a period of years if wheat is grown in mixed stands with corn cockle *(Agrostemma githago)*.[40,41] Sterile corn-cockle seedlings were found to stimulate the growth of sterile wheat seedlings on an agar medium, thus demonstrating that an exuded chemical was involved. Three compounds stimulatory to the growth of wheat seedlings—agrostemmin, allantoin, and gibberellin— were isolated from corn-cockle seeds. Gajić and several colleagues reported that application of agrostemmin to wheat fields at the rate of 1.2 grams per hectare increased yields of wheat grain in both fertilized and unfertilized areas.[42]

Fig. 6. Zonation of species around sunflower *(Helianthus annuus)* in a field plot near Norman, Oklahoma. *S.* sunflower. *Bromus japonicus* is growing near the sunflowers; *Erigeron canadensis, Rudbeckia serotina,* and *Haplopappus ciliatus* are in the zone away from the sunflowers. Reproduced from E. L. Rice, *Allelopathy,* courtesy of Academic Press.

ALLELOPATHIC EFFECTS OF CROP PLANTS ON WEEDS

Fortunately, chemical interactions between crop plants and weeds do not always favor the weeds. Numerous instances have been reported of crop plants producing substances inhibitory to weeds. Both thin and dense field stands of Kentucky-31 fescue *(Festuca arundinacea)* are often relatively free of weeds. E. J. Peters has demonstrated that toxic subtances are exuded from the roots of fescue, and these chemicals inhibited the growth of black mustard *(Brassica nigra)* and trefoil *(Lotus corniculatus)* in tests.[43]

N. N. Dzubenko and N. I. Petrenko demonstrated that root exudates of lupine *(Lupinus albus)* and corn inhibited the growth of lamb's-quarters *(Chenopodium album)* and pigweed *(Amaranthus retroflexus)*.[44] Wheat, oats, peas, and buckwheat *(Fagopyrum sagittatum)* also suppress growth and accumulation of biomass in lamb's-quarters.[45] S. A. Markova found that oats suppress the growth of wormseed mustard *(Erysimum cheiranthoides),* at least in part because of the production of a toxin.[46] N. I. Prutenskaya reported that crunchweed *(Sinapis arvensis)* was strongly inhibited by wheat, rye, and barley, but was stimulated by broom-corn millet *(Panicum milliaceum)*.[47]

Allelopathy in Biological Weed Control.

The research discussed just above certainly suggests the possibility of breeding crop plants that control some of the more important weeds in a given area through production of allelopathic compounds. Geneticists have been breeding crop plants for many years that are resistant to many diseases. Many of the compounds produced by plants that inhibit the growth of pathogens are the same as or related to the chemicals that act as allelopathic agents.

Rhizobitoxine, a compound produced by some strains

of the bacterium *Rhizobium japonicum,* is an effective herbicide in amounts as low as 3 ounces per acre. Agrostemmin, one of the allelopathic agents produced by corn cockle, decreased numbers of weeds and increased yields of desirable grasses when applied in the amount of 1.2 grams per hectare to selected pastures in Yugoslavia.[48]

A. R. Putnam and W. B. Duke screened 526 accessions of cucumber and 12 accessions of eight related species from forty-one nations of origin for allelopathic activity against a weedy forb, white mustard *(Brassica hirta),* and a grass, broom-corn millet.[49] One accession inhibited the growth of test plants by 87 percent, and twenty-five accessions inhibited growth by 50 percent or more. The researchers concluded that incorporation of an allelopathic character into a crop cultivar could give the plant a competitive advantage over certain important weeds.

P. K. Fay and W. B. Duke screened three thousand accessions of oat *(Avena* spp.) germ plasm, in the United States Department of Agriculture Collection for their ability to exude scopoletin, a naturally occurring compound shown to have growth-inhibiting properties.[50] Twenty-five accessions exuded more of the blue-flourescing materials characteristic of scopoletin from their roots than a standard oat cultivar, Garry oats. Four accessions exuded up to three times as much scopoletin as Garry oats. When one of the four was grown in sand culture for sixteen days with a wild mustard *(Brassica kaber),* the growth of the mustard was significantly less than when the weed was grown with Garry oats. The mustard plants grown in close association with the allelopathic accession exhibited the severe chlorosis, stunting, and twisting indicative of chemical effects rather than simple competition.

M. A. Panchuk, N. I. Prutenskaya, and A. M. Grodzinsky worked on the transfer of genes for toxin production in hybrids of wheatgrass *(Agropyron glaucum)* and Lutescence 329 wheat.[51,52] They found that water extracts of wheatgrass residues were more toxic than wheat-residue extracts, when tested against the seed germination of radish and the root growth of peppergrass *(Lepidium sativum)*. The first-generation hybrids exhibited chiefly wheatgrass characteristics and manifested high inhibitory activity. The inhibitory activity of the other hybrids studied was intermediate. This is certainly just a start on the genetics of allelopathic agents, and much research will have to be done to realize the tremendous potential for weed control by the allelopathic action of crop plants.

Underground Bridges Between Plants. Root exudates received a great deal of attention in this chapter. The term implies that the chemical inhibitors move out of the roots and into the soil, where they can be taken up again by the roots of other plants. This has been demonstrated to happen, of course, but there is a large body of evidence indicating that many vascular plants often do not grow as distinct individuals, but are interconnected with other individuals and even different species.[53] The bridges between the plants are formed by natural root or stem grafts, mycorrhizal fungi, or haustorial connections of parasitic vascular plants.

B. F. Graham and F. H. Bormann listed almost two hundred species of angiosperms and gymnosperms that have been reported to form natural root grafts.[54] There is no doubt that many kinds of chemical compounds can pass through the grafts, because even the spores of pathogenic fungi, such as the fungus that causes Dutch elm disease, can be transmitted to other individuals through root grafts.[55]

The flowering plant called pinesap or Indian pipe, *Monotropa hypopitys,* is well supplied with mycorrhizal fungi, and these fungi are shared with pine and spruce trees. Moreover, compounds pass from the trees to the pinesap plant through the fungal bridges.[56] Mycorrhizal sharing has also been demonstrated between several of the nongreen orchids and various forest trees.[53]

Mycorrhizal fungi are apparently essential to the growth of the majority of the world's commercially important plants.[57] Frank Woods and K. Brock have suggested, moreover, that mycorrhizal fungi may be mutually shared by the root systems of many forest trees. If this is so, allelopathic compounds could readily move from plant to plant through the fungal bridges. These scientists applied either radioactive phosphorus (^{32}P) or radioactive calcium (^{45}Ca) in aqueous solutions to freshly cut stumps of young red maples *(Acer rubrum)* in a mixed hardwood forest stand.[58] Subsequently they took foliage samples from all of the trees within 24 feet of the treated ones and analyzed these for the presence of the radioactive element. They found that the radioactive material was transferred into nineteen different species of shrubs and trees surrounding the donor trees. They pointed out that the material could have moved from the donor plants to the others (1) through root grafts, (2) by root exudation and uptake, and (3) through mutually shared mycorrhizal fungi. On the basis of previous studies they suggested that the trees were too young to have root grafts. Therefore the results were probably due to exudation or fungal bridges. Woods and Brock pointed out that the exchange rates were such that within a period of several weeks every woody plant in the local community would have acquired at least a few of the radioactive ions originally introduced to the donor. Mycorrhizae are present in herbaceous plants also, but less is known about their overall significance in such plants.

ALLELOPATHY IN PLANT DISEASES

Allelopathic agents appear to be involved in several processes associated with plant disease: development and morphogenesis of pathogens, antagonism to pathogens of nonhost organisms, development of disease symptoms, and host-plant resistance to pathogens.[59] The promotion of infection by allelopathic compounds in the environment of the potential host plant should be added to the list. These are all rather complex topics, and much research has been done on most of them. Therefore only a few examples are given below as illustrations.

The spores of most parasitic fungi are formed in dense populations in or on infected host tissue. For several reasons they usually remain ungerminated while located at their site of production.[59] One factor is the production by the spores of fungistatic agents which they excrete into surrounding water. These self-inhibitors assure the dispersal of viable ungerminated spores.

Germination of the reproductive structures of fungi is often promoted subsequently by chemicals produced by potential host plants. For example, a volatile principle evolved from onion and leek seedlings causes the germination of the dormant sclerotia (reproductive structures) of a fungus called *Sclerotium cepivorum.* Volatile compounds released from germinating seeds have marked effects on many microorganisms. G. Stotsky and S. Schenck surveyed several plant species for the production of such compounds.[60] R. N. Allen and F. J. Newhook have demonstrated that ethyl alcohol suppresses the chemotactic response (the movement in relation to a chemical) of the disease-causing fungus *Phytophthora cinnamomi.*[61] Later ethyl alcohol was found in the vicinity of lupine roots. The concentrations of the alcohol were commonly as great as those that previously had been shown to attract zoospores of *Phytophthora.*

The survival of most parasites depends on the formation of resting propagules, because such organisms have to spend long periods of time away from their host plants. There is considerable evidence that formation of these propagules may be conditioned by allelopathic chemicals.

Most propagules of pathogens do not survive to infect suitable host organisms because of many adversities, such as the antagonism of other organisms. Allelopathy is one kind of antagonism. Researchers have been concerned with the production of antibiotics, both by other pathogens and by microorganisms that live on dead organic matter. Since antibiotics are nonnutritional substances produced by one microorganism that are effective against another microorganism, they are allelopathic agents. Many references are cited by A. A. Bell for the reader who wishes more details on this subject.[59]

Fungistasis is a widespread phenomenon to which many causes have been attributed, and for which many factors have been investigated. Like the sources of allelopathic agents in soil, its origins are not always known. R. R. Mishra and K. K. Pandey concluded from their data that the inhibition of spore germination has a biological origin.[62,63] A great many of the compounds that are known to inhibit spore germination and fungal growth enter the soil as products of higher plants. It is certainly possible therefore that some of those compounds may be at least partially responsible for fungistasis.

Much evidence has been presented concerning production by higher plants of compounds antagonistic to the root-rot fungus *Poria weirii*.[64–67] C. Y. Li and his coworkers found that many phenolics and other compounds that repeatedly have been isolated from soil inhibit the growth of *Poria weirii*. They found that the combination of phenolic compounds found in the roots of the *Poria*-resistant red alder *(Alnus rubra)* inhibited the growth of *Poria* also. Many species of plants that form an understory in red-alder stands were demonstrated to produce several of the same phenolic compounds inhibitory to *Poria*.

The antagonistic activity of certain microorganisms can be used in biological control of plant diseases.[68–70] *Fomes annosus* is a fungal root pathogen of loblolly and slash pine. Its spores germinate on cut stumps, grow down through the stump, and infect living roots through natural root grafts between plants. Apparently inoculation of cut stumps with spores of another fungus, *Peniophora gigantea,* can prevent the disease from occurring because *Peniophora* prevents entry of *Fomes annosus.*[71] A. F. Yakhontov found that water-soluble excretions of many weeds and crop plants inhibit the fungus that causes the disease phylloxera in grape plants.[72] He observed also that there were different degrees of infection by phylloxera in areas where different grasses grew alongside the grape plants in the vineyards.

A tremendous amount of research has been done on the phytotoxins produced by pathogens that cause various disease symptoms to appear in the host plants. The reader is referred to the Notes for details on this fascinating and important subject.[59,73,74]

The compounds involved in the resistance of host plants to pathogens are divided into two categories: (1) secondary compounds that are generally present in the host but may increase subsequent to infection and (2) phytoalexins, which are new compounds formed only after infection. In 1963, G. L. Farkas and Z. Kiraly gave a thorough review of secondary compounds involved in the resistance of cereal grains to fungal pests.[75] More recent are the general reviews on the roles of secondary compounds by A. A. Bell[59,76] and T. Swain.[77] Much research has been done on phytoalexins in this decade, and the reader is referred to the relatively recent reviews of the subject by J. Kuć,[78] Bell,[59,76] and R. K. S. Wood and A. Graniti.[79]

Infection Promoted by Allelopathic Compounds

As early as 1948, V. W. Cochrane suggested that some root rots are initiated by the direct toxic action of plant residues.[80]

Z. A. Patrick and his colleagues have obtained considerable evidence to support that suggestion. In field studies in California they observed that discolored or sunken lesions were often present on the roots of lettuce or spinach plants where the roots grew in contact with or in close proximity to fragments of plant residues.[7] When isolations of microorganisms were made from the lesions, no known primary pathogen was consistently obtained, and the microorganisms most frequently found were common soil saprophytes (organisms that generally grow on dead organic matter). They concluded that the toxins produced by the decomposing residues conditioned roots to invasion by various low-grade pathogens.

T. A. Toussoun and Z. A. Patrick reported that the bean root rot caused by the fungus *Fusarium solani* f. *phaseoli* increased greatly if the bean roots were exposed to toxic extracts from decomposing plant residues before inoculation with the pathogen.[81] Patrick and L. W. Koch found that the extent and severity of black root rot of tobacco (a disease caused by another fungus) were much greater when the tobacco roots were exposed to toxic extracts of decomposing plant residues prior to inoculation.[82] Additionally, they reported that the pathogen was equally destructive to susceptible and normally resistant varieties of tobacco if their roots were treated with toxic extracts.

CHAPTER 4

Plant-Nematode Chemical Interactions in Agriculture

THE nematodes are commonly called threadworms, eel-worms, or roundworms. Their habitats are more varied than those of any other group of animals except the arthropods. The common nematodes that live in soil and water are usually too small to be seen with the naked eye; others, particularly some marine and parasitic species, are easily within the range of vision. Some are but 1/125 of an inch long, whereas others grow to relatively enormous dimensions. One species that lives in the kidneys of dogs and other mammals may be a yard long and attains the thickness of the little finger. Nematodes obtain their sustenance in many ways: some live on dead organic matter; some are herbivores; some are carnivores that prey upon other microscopic animals, including other nematodes; many are parasites on animals; and many are parasites on plants. N. A. Cobb estimated that as many as three billion nematodes may live on one acre of soil, with most of those in the top three inches.[1]

In plants nematodes attack the root, bulb, stem, leaf, flower, fruit, or seed; in animals they are found in the eye, mouth, tongue, digestive tract, lungs, liver, body cavity, muscle, or joint. In fact, almost any location appears to be suitable for at least a few species of nematodes. The huge numbers of nematodes that are either predatory or parasitic on crop plants make them extremely important in agriculture and worthy of a separate chapter in relation to plant-animal chemical interactions.

Marigolds. The marigold has been an important medicinal plant since the first century A.D.[2] As described in Chapter 2, Coles reported in 1657 that the juice of marigold flowers dropped into the ears kills "worms."[3] Much later the marigold was found to be valuable for pest control in agriculture. J. Tyler in 1938 and G. Steiner in 1941 reported that many marigold species (*Tagetes* spp.) are resistant to the infestations of many species of nematodes.[4,5]

In 1957, M. Oostenbrink and his coworkers[6] reported that the presence of marigold appears to suppress the population of some nematodes in soil, and J. H. Uhlenbroek and J. D. Bijloo[7] reported in 1958 that a preparation extracted from *Tagetes nana* was highly destructive to several nematodes. One powerful nematocide identified was α-terthienyl. T. Visser and M. K. Vythilingam demonstrated in 1959 that interplanting either of two species of marigold (*T. erecta* or *T. patula*) with tea plants controlled meadow eelworm *(Pratylenchus coffeae)* and root-knot eelworm *(Meloidogyne javanica)* on tea more effectively than fallowing the land.[8]

A. M. Omidvar reported in 1961 that root diffusates of three marigold species had no effect on the hatching of the cysts of potato-root eelworm *(Heterodera rostochiensis),* whereas diffusates of potato, tomato, and eggplant stimulated the hatching of the cysts.[9] Those three crops are hosts of this nematode. Subsequently, Omidvar reported that, if marigold plants were grown in soil infested with potato-root eelworm cysts, the eelworm was slightly reduced "possibly due to weak nematicidal secretion."[10] J. J. Hesling and his coworkers also tested root diffusates from marigold *(T. minuta)* against the hatching of potato-root eelworm cysts. They found no effect, but a dilute diffusate from potato roots stimulated germination eightyfold.[11]

R. A. C. Daulton and R. F. Curtis grew three species of marigold *(T. erecta, T. patula, T. minuta)* in soil that was infested by root-knot eelworm *(Meloidogyne javanica)*. They found that each marigold species reduced the eelworm population to low levels in 42 to 70 days.[12] Root-knot larvae failed to penetrate the marigold roots in any appreciable numbers, and those that entered did not develop beyond the infective second-stage larval form. Spectacular nematode control was obtained by interplanting alternate crops of tomato with marigolds. *Tagetes minuta* is an abundant weed in most parts of southern Rhodesia, especially on the light, sandy tobacco-growing soils. The root-knot eelworm *M. javanica* is by far the predominant root-knot nematode species in Rhodesian soils, and it is the cause of large-scale crop losses annually. The economic potential of these findings is obvious.

Strikingly successful nematode control has been obtained in tobacco rotations in Rhodesia by interplanting alternate crops of tobacco with various marigold species.[13] Equally spectacular results have been reported in controlling the *Pratylenchus* nematode on tobacco in Germany by alternating crops of marigolds.

Tagetes minuta was found to be resistant to most root-knot nematodes of the genus *Meloidogyne* that were tested, and this marigold was found to give excellent control of five genera of nematodes in field tests in Tifton sandy loam in Georgia.[14] The tomato plants grown after marigolds were of better quality than those grown after beggarweed *(Desmodium tortuosum)*, sudan grass, or millet. No marketable plants were produced after millet, and only 6,000 and 28,000 plants per acre were produced after beggarweed and sudan grass, respectively, compared with 302,000 marketable plants per acre after marigolds. Tomato plant growth was also more uniform after marigolds than after all of the other crops except rattlebox *(Crotalaria).*

Root-knot nematodes *(Meloidogyne incognita)* develop slowly in marigold *(T. patula)* roots, and giant cells and nematodes degenerate in many infection sites according to R. W. Hackney and O. J. Dickerson.[15] These workers found that marigold is very effective in reducing the numbers of this nematode in tomato roots. Tomatoes grown in soil previously planted to marigold had a root-knot index of 1.0, whereas the index was 3.9 in soil previously planted to tomatoes. The root-knot indices are as follows:

$$0 = 0\% \text{ galled roots}$$
$$1 = 0\text{-}25\% \text{ galled roots}$$
$$2 = 26\text{-}50\% \text{ galled roots}$$
$$3 = 51\text{-}75\% \text{ galled roots}$$
$$4 = 76\text{-}100\% \text{ galled roots}$$

The index value was increased 0.5 when egg masses were observed. Root populations of another parasitic eelworm, *Pratylenchus alleni,* were significantly less in marigolds than in tomatoes; and root populations of *M. incognita* and *P. alleni* were significantly less where tomatoes were simultaneously cultivated with marigolds than where they were cultivated alone.

In India several species of marigold are grown between beds of vegetables belonging to the potato family, and the direction of the marigold beds is changed each year. Farmers have continued this practice from time immemorial without knowing its exact significance. A. M. Khan found that, where cultures of African marigold *(T. erecta)* were mixed with four varieties of tomato, the root-knot index was markedly reduced. The marigolds also decreased the numbers of five other nematode genera in the soil, as well as the root-knot nematode larvae *(Meloidogyne).*[16] The weights of the tomatoes also increased markedly. Similar results were obtained with okra *(Hibiscus esculentus)*.

Khan reported that, of fifteen varieties of cultivated tomatoes tested for root-knot nematodes *(M. incognita)*, all were susceptible in some degree, but resistance varied. One wild species tested, *Lycopersicum pimpinellifolium*, was resistant, suggesting the presence of a toxic chemical. Khan also tested twenty cucurbits from among six genera, and all were moderately to highly susceptible to root-knot nematodes. He stated, however, that *Cucumis* sp. var. Bikaner and *Cucurbita moschata* var. Jaipuri were tolerant to infection by root knot.

In soil at Aligarh, India, interplanting of African marigold with eggplant and chilis markedly lowered the populations of some six genera of nematodes, compared with the numbers present when the chilis and eggplants were planted alone.[17] Marigold also reduced the populations of plant-parasitic nematodes infesting cabbage and cauliflower. The vegetative growth and the dry weight of all the vegetable crops tested were considerably increased when they were grown with marigold. The inhibitory effects of marigold on nematode populations may be caused by toxic chemicals that are present in its root exudate, according to M. M. Alam and his coworkers.[18]

P. M. Miller and J. F. Ahrens studied differences in nematode populations and plant growth for three years after crops of marigolds *(T. patula)*, rye, buckwheat, pigweed, and crabgrass in 1963 and after fumigation in the spring of 1964 with six gallons per acre of ethylene dibromide.[19] Marigolds suppressed populations of the eelworm *Pratylenchus penetrans* for three years and *Tylenchorhynchus claytoni* for one year. Fumigation suppressed *P. penetrans* for two years and *T. claytoni* for three years. Rye, buckwheat, pigweed, crabgrass, and several other weeds in the plots were good hosts for both parasitic nematodes. Tobacco grew better in the fumigated plots in 1964, and where marigolds had grown in 1963, than it did where rye, crabgrass, pigweed, and buckwheat had grown in 1963. Marigolds improved the growth of privet for two years, whereas ethylene dibromide was toxic to privet. Tobacco quality was highest in 1964 in marigold and fumigated plots. In greenhouse tests tobacco, tomato, potato, petunia, zinnia, balsam *(Impatiens),* stock, and aster grew better in fumigated soil or soil where marigolds had grown than in soil containing a natural infestation of parasitic nematodes.

R. Winoto-Suatmadji found that various species of marigolds markedly suppressed several species of the eelworm *Pratylenchus* in tube cultures and in field trials.[20] Also the marigolds were generally as effective or better than fallow in suppressing several species of the root-knot nematode *Meloidogyne.* Several species of another parasitic nematode, *Tylenchorhynchus,* were suppressed by *T. patula* in different soils. *Tylenchorhynchus dubius* was suppressed better and more rapidly by marigolds than by fallowing the land. Marigold soil was distinctly nematocidal for some days after the roots were removed. Root extracts of marigolds were found to contain a nematocidal chemical or chemicals. Winoto-Suatmadji suggested that these chemicals might be the thiophenes previously isolated and identified from marigold roots.

Simultaneous culture of marigolds with a main crop appears to be effective around and between trees and woody ornamentals. It may prove beneficial in many situations, since sowing marigolds 60 centimeters apart completely suppresses nematodes.[20] Autumn sowing of marigolds after a main crop, and planting marigolds between the rows of a main crop, may be promising under certain conditions. Marigolds promoted the growth of apple seedlings in soil with *P. penetrans* to 167 percent of growth in fallow infested soil, but marigolds decreased apple-seedling growth in uninfested soil. A direct mulch with a natural dosage of marigold roots suppressed *P. penetrans* much better than other mulches or fallowing the land. Marigold-leaf mulches were also effective in suppressing *P. penetrans,* as are mulches of some genera of another tribe of the composite family, the Heleniae (sneeze-weeds).

Four other plant genera—the rattlebox *(Crotalaria),* chrysanthemum *(Chrysanthemum),* castor bean *(Ricinus),* and margosa *(Azadirachta)*—have been shown to be effective in reducing soil nematode populations and crop infestations in several localities. Again it is noteworthy that at least two of these plants, the chrysanthemum and the castor bean, have been used medicinally since at least the first century A.D.

Rattlebox. Root-knot nematodes were controlled when rattlebox *(C. spectabilis)* was interplanted in peach orchards for a period of two years, according to C. W. McBeth and A. L. Taylor.[21] Actually, rattlebox was planted in the summer, and oats, which are resistant to root-knot nematodes, were planted in the winter, and both were turned under as green manure. The growth and yield of the Elberta peach trees were greatly increased over the susceptible cover crops that were grown as controls. Whippoorwill cowpeas were grown in the summer as a control, and Austrian winter peas in the winter, with both turned under as green manure.

Twenty selections of rattlebox were evaluated in Florida by J. J. Oschse and W. S. Brewton. Seven were found effectively to reduce the numbers of parasitic nematodes in soil.[22] Those seven were distributed among three rattlebox species, *C. pumila, C. mucronata,* and *C. brevifolia.*

Two rattlebox species, *C. usaramoensis* and *C. anagyroides,* were tested by T. Visser and M. K. Vythilingam for parasitic nematode control on tea plants, and both reduced meadow eelworm and root-knot eelworm on tea more effectively than fallowing the land.[8] *C. mucronata* reduced numbers of *Pratylenchus zeae* and *P. brachyurus* in soil in a series of tests by B. Y. Endo.[23]

Rattlebox, *C. spectabilis,* demonstrated a high degree of resistance and low root-knot galling against at least five species of the root-knot nematode *Meloidogyne,* according to J. M. Good and his colleagues.[14] Field tests demonstrated that this rattlebox is very effective in reducing soil populations of a wide range of parasitic nematodes, including species in at least six different genera. The tests were made in a sandy loam soil in Georgia in the United States.

Chrysanthemums. Investigations as early as 1930 demonstrated that some varieties of chrysanthemum are resistant to the chrysanthemum eelworm, *Aphelenchoides ritzemabosi.* Unfortunately, uninfested controls were not tested, and the sizes of the initial infestations on the plants were not known. J. J. Hesling and H. R. Wallace made thirteen varieties of chrysanthemum eelworm-free by hot-water treatments and tested them for resistance to chrysanthemum eelworm by artificially infesting them with known numbers of eelworms in different stages of development.[24] Most of the cuttings in all of the chrysanthemum varieties eventually showed leaf browning or distortion, indicating no differences in susceptibility at this early stage of growth. Mature plants, on the other hand, showed marked differences in susceptibility to damage between different varieties. Susceptibility was not determined by stomatal size or frequency, the density of epidermal hairs on leaves or stems, the size of mesophyll air spaces, or the thickness of leaf cell walls. Subsequently the leaves of resistant varieties of chrysanthemum were found to brown quickly when infested with the chrysanthemum eelworm.[25] In those species there is no eelworm multiplication, and thus the infestation is isolated and does not spread. The infested leaves of resistant and susceptible varieties apparently brown at different rates because of the responses of the gravid female worms. In the leaves of resistant varieties the females move about, piercing hundreds of cells, which subsequently turn brown. In susceptible varieties the females do not move; they pierce few cells and lay many eggs. The difference in behavior was attributed to the absence of a nutritional factor in the leaves of resistant varieties, but no evidence was presented to support this. Some antifeedant or other chemical factor could be responsible instead.

R. W. Hackney and O. J. Dickerson found that populations of the root-knot nematode *Meloidogyne incognita* were significantly less in the roots and soil of chrysanthemums (Escapade) than in tomato roots and soil. Moreover, root populations of *M. incognita* and *Pratylenchus alleni* were significantly less in tomatoes that had simultaneously been cultivated with chrysanthemums than in tomatoes that were cultivated alone.

Castor Beans. Root and soil populations of *M. incognita* were also found to be significantly lower among castor beans than among tomatoes.[15] Aborted giant cells and dead root-knot nematode larvae and females were observed in castor-bean roots, but not in tomato roots. Moreover, root populations of *M. incognita* and *P. alleni* were significantly lower in tomatoes cultivated with castor beans than in tomatoes cultivated alone. Thin-layer and column chromatography of alcohol extracts from castor bean revealed no nematocidal thiophene derivatives such as are present in marigold roots. These results suggest the desirability of more investigations concerning the possible use of castor beans in the biological control of at least some nematodes.

Margosa. M. M. Alam and coworkers reported that root exudates of margosa *(Azadirachta indica)* are toxic both to nematodes and larval hatch of nematodes.[18] Mixed cropping of margosa with several crop plants in Aligarh, India, demonstrated that margosa reduced the populations of at least six genera of plant-parasitic nematodes on tomato, eggplant, cabbage, and cauliflower.[17] Moreover, the growth of all the vegetable crops tested was promoted when cultivated along with margosa.

Asparagus. Root exudates of asparagus *(Asparagus officinalis* var. *altilis)* were demonstrated by R. A. Rohde and W. R. Jenkens to contain nematocidal compounds that reduce the populations of plant-parasitic nematodes in soil.[26] Apparently this finding has never been pursued further, and it appears it would be highly desirable to do so.

Other Crop Plants. Many additional crop plants, including numerous genera of pasture grasses, have been investigated from the viewpoint of using some of them for control of parasitic nematodes. As one might expect, results have been mixed (see Notes 8, 11, 14, 19, 21, 23). Oats appear to hold some promise.[21,23] Guatemala grass *(Tripsacum laxum)* was effective in reducing meadow-eelworm infestations when intergrown with tea plants.[8]

Extracts of pumpkin, cucumber, garlic, and hundreds of other plants have been found to be toxic to nematode parasites in animals. S. I. Husain and A. Masood observed that a great many medicinal plants destroy or expel parasitic worms, and they tested eleven such species against larval hatching of the root-knot nematode *Meloidogyne incognita.*[27] After making 10-percent aqueous solutions of dried powder from the leaves, seeds, and flowers of all eleven species, they found that all of the extracts inhibited larval hatching. In all cases the hatching percentage increased with an increase in dilution of the extracts but, even in a thousandfold dilution, hatching was much lower than in controls. Leaf extracts of Indian aloe *(Aloe barbadens),* margosa, and wormseed

(Chenopodium anthelminticum) were strikingly inhibitory. Of these apparently only margosa has been field-tested for control of parasitic nematodes in crop plants.

Thus it appears that biological control of plant-parasitic nematodes by plant-nematode chemical interactions is being practiced in many parts of the world. Moreover, the future looks bright for the extension of this control by using promising new plants.

CHAPTER 5

Plant-Insect and Insect-Insect
Chemical Interactions in Agriculture

AN INSECT is a segmented animal with a tough outer integument and jointed limbs. Its body is divided into head, thorax, and abdomen, and it breathes through air tubes. The head is the sensory and feeding center, bearing the mouthparts and a single pair of antennae. Compound eyes are usually present, and often simple eyes as well. The thorax is the center of locomotion, bearing three pairs of legs and usually two pairs of wings. The abdomen is the metabolic and reproductive center, containing the gonads, the organs of digestion and excretion, and usually some special structures used in copulation and egg laying.

Most insects go through definite developmental stages. In some species the infant looks like a small copy of the adult, and it grows larger by a series of molts until it reaches full size. The immature, nonreproductive forms are known as nymphs. Almost 90 percent of all insect species, however, undergo a complete metamorphosis, so that the adult is completely different from the immature form. The immature eating and feeding forms may be correctly referred to as larvae, although they are commonly called caterpillars, grubs, or maggots, depending on the species. After the larval period the insect undergoes a complete metamorphosis and enters an outwardly quiescent pupal stage, in which a complete change in form occurs. The adult emerges from the pupa.

Larvae are concerned almost entirely with eating and growth. Some eat only a few specific kinds of food (oligophagous), whereas others eat many kinds of food (polyphagous). As they grow, they molt a characteristic number of times, the number depending on the species. The stages between molts are known as instars. When the larva is full-grown, it molts once more to form the pupa. In the early 1930s, Vincent Wigglesworth proved that metamorphosis in insects is controlled by hormones.[1] Pertinent aspects of the process will be discussed in Chapter 6 in connection with insect control in crops.

The number of named species of insects exceeds the number of all the other animals together. It approximates 800,000 species. Usually several thousand new species are described each year, and probably the insect species remaining to be named exceed all of the known kinds.[2] The order of beetles, the Coleoptera, alone comprises over 330,000 named species, and the weevil family, the Curculionidae, includes over 60,000 known species.

No other single class of animals has so thoroughly invaded and colonized the globe as the Insecta. Every species of flowering plant provides food for one or more kinds of insects, and this is probably true of all other plants also. Decomposing organic matter attracts and supports many thousands of species, and many insects are parasites on or in the bodies of other insects or of some very different order of animals, such as the vertebrates. Soil and fresh water also support an extensive insect fauna. Extreme temperatures are not impassable barriers. Some insects can withstand temperatures of $-50°C$, while others live in hot springs at over $40°C$ or in deserts, where the midday surface temperatures may be twenty degrees higher.[2] The British entomologist C. B. Williams has estimated that 10^{18} (a billion billion) individual insects are alive at every instant. This amounts to 10 billion for every square kilometer of land surface and 200 million for every human being.[3] It is not surprising therefore that insects are the leading consumers of plants and the most important pests of agriculture everywhere in the world.

CHEMICAL EFFECTS OF THE HOST PLANT AND OTHER INSECTS ON INSECT BEHAVIOR AND REPRODUCTION

The reactions of an insect to its chemical environment are delayed or immediate. Delayed reactions include symptoms of toxicity, growth, hormonal changes, and reproductive differentiation.[4] Immediate reactions involve the interaction of chemicals with the insect's external sense organs, causing many behavioral manifestations. Some of the overt behavioral responses to chemical stimulation of the sense organs are initiation or termination of movement, feeding, courtship, copulation, oviposition, grooming, defense of territory, aggressive and defensive actions, nest construction, and different phases of parental care.

Up to 1960 there was considerable confusion in the terminology used to designate the chemicals that evoke insect responses. Therefore, V. G. Dethier and some colleagues suggested the following terms to describe both animal-animal and plant-animal chemical interactions:

(1) An *attractant* causes animals to make oriented movements toward its source.

(2) An *arrestant* causes animals to aggregate.

(3) A *stimulant* elicits an activity, such as feeding or oviposition.

(4) A *deterrent* inhibits such an activity.

(5) A *repellent* causes an animal to make oriented movements away from its source.[5]

Although the terms were suggested for insects, they apply to other animals as well. The senses primarily involved in the detection of plant chemicals are smell and taste, but sight is frequently involved also.

It is important to understand the following principles: (1) The same compound may have multiple effects on behavior. (2) Any given effect may be elicited by more than one chemical. (3) Movements involving orientation can be triggered by concentration gradients or by air currents carrying a chemical that may not have to be present as a gradient.

Chemicals of plant origin may fit into any of the categories suggested by Dethier and his colleagues. Although plant-produced attractants sometimes become arrestants in higher concentrations near their source, it appears that most arrestants are of animal origin and are therefore pheromones. Chemicals in insect pollination will be discussed later.

P. A. Hedin and his colleagues have provided long lists of plant-produced feeding stimulants, attractants, feeding deterrents, and repellents, along with the names of affected insects and literature references.[6] Their article is an excellent supplement to the information presented in this chapter.

Attractants

There seems little doubt that plant-eating insects find appropriate host plants through their senses of sight and smell. In either case chemicals are involved, and generally the chemicals are different from those used in nutrition. No doubt this is also true of insects that are predators or parasites of animals. Much research has been done on feeding stimulants and deterrents, but there has been relatively little on attractants, especially those that draw insects to crop plants.

In 1910, E. Verschaffelt determined experimentally that

cabbage-butterfly larvae *(Pieris brassicae* and *P. rapae)* are attracted by the various mustard oils contained in the host plants. Verschaffelt theorized that the larvae smell the odors from the mustard oils before they begin to eat the food.[7]

It is easy to determine that silkworms can smell because, if mulberry leaves are placed near them where they can neither see nor touch the leaves, the silkworms move their heads, work their mouthparts, and begin crawling toward the leaves. N. E. McIndoo also tested the responses of tent caterpillars, fall webworms, tussock-moth larvae, armyworms, and the larvae of the black swallowtail butterfly *(Papilio polyxenes)* to the odors of numerous plants and to protruded thoracic glands of *Papilio* larvae. He found that the larvae usually responded to the odors from the various sources.[8]

Vincent Dethier demonstrated conclusively in 1937 that larvae of the monarch butterfly *(Danaus plexippus)* locate their milkweed *(Asclepias)* hosts by odor. He did not identify the chemical attractant, however.[9]

Cabbage butterflies and the diamondback moth *(Plutella maculipennis)* are oligophagous insects specific to cruciferous plants. Such plants characteristically contain mustard-oil glucosides and usually only small amounts of mustard oil or none at all. A specific enzyme, myrosin, is also present in the mustard plants, but in different compartments from those that contain the glucosides. Slight injury can bring these substances together, producing mustard oils plus other compounds. A. J. Thorsteinson found that the mustard oils are feeding deterrents to the cabbage butterflies and the diamondback moth but suggested that minute amounts may serve as attractants.[10]

The rice weevil *(Sitophilus zeamais)* is an important stored-products insect that causes serious losses in stored grain. The weevils were found to migrate actively toward an ether extract of the rice grains, but migration occurred only during the first ten minutes toward a water-soluble fraction.[11] Characterization of the active fraction indicated that it was stable in heat (150°C) for one hour, in ultraviolet light for two hours, in aqueous potassium hydroxide for two weeks, and after refluxing for one hour in a 3-percent solution of alcoholic potassium hydroxide. It was not volatile at 150°C and was acidic. Volatile components were detected also from nitrogen-gas aeration of the rice grains and from trapping in dry-ice and acetone. Neither compound nor mixture was identified.

The rice-stem borer *(Chilo suppressalis)* is one of the most destructive pests on rice plants in Japan and East Asian countries. In early summer the larvae pupate after hibernating in the rice straw and stubble. Then the moths emerge. After mating, the females migrate to the paddy fields and lay their eggs on the rice plants.[12] The newly hatched larvae immediately bore into the stems, where the first generation pupate and develop into moths in the late summer. The second-generation larvae bore into the stems and pass the winter in the full-grown larval stage inside the infested stems or stubble. Since there are varietal differences in the susceptibility of rice to the attacks of the borer, experiments were undertaken to determine whether an attractant is present and whether it varies in activity in different varieties. The active attractant was identified as p-methyl acetophenone, and it was given the common name "oryzanone" after the name of the rice plant, *Oryza*.

It has been known for many years that the volatile oil from cotton plants is attractive to the cotton boll weevil *(Anthonomus grandis)*. J. H. Tumlinson and his colleagues fractionated and examined a steam distillate of several tons of cotton buds and whole plants and isolated six volatile terpenoid compounds and a major volatile alcohol, ß-bisabolol, all of which had attractant activity.[13] Of the attractant activity 60 to 80 percent appeared to be caused by the alcohol. All the terpenoids were identified except for two sesquiterpenoids (15-carbon atoms). Thus several compounds are involved, and they appear to have a synergistic effect on each other when combined.

Fruit-piercing moths—such as the akebia-leaf-like moth *(Adris amurensis),* the reddish oraesia *(Oraesia excavata),* the small oraesia *(Oraesia emarginata),* and others—are important juice-sucking moths on many orchard fruits in Japan and many tropical countries.[12] In the evenings these moths fly long distances from mountain shrubs to orchards, where they suck juices by piercing the fruits with a tough proboscis. In the early morning they fly back to their resting places.

Large numbers of the fruit-piercing moths are attracted to traps containing ripe fruit that are located away from the orchards. This suggested that such fruits contain an attractant. In one experiment dry air was passed through a drum containing ripe grapes.[12] The air was then chilled, and a condensate was collected in a flask immersed in an alcohol-dry-ice bath at $-70°C$. Subsequent fractionation of the condensate isolated a very active chemical attractant, which T. Saito and K. Munakata did not succeed in identifying.

Food plants of the vegetable weevil *Listroderes obliquus* include 178 species or varieties in thirty-four plant families. It was generally accepted for many years that such polyphagous insects do not respond to specific chemical stimuli but simply sample randomly until an acceptable host is encountered. In a survey and classification of food-plant records of *L. obliquus,* it was found that the Compositae ranked first in the number of species represented, followed by the Cruciferae and the Umbelliferae.[14] A greater number of food-plant species in certain plant families does not necessarily indicate that the insect prefers those plants among the many recorded. Nevertheless, it is true that many plants in the Cruciferae and Umbelliferae are severely damaged by the vegetable weevil. Therefore the attraction of the volatile mustard oils, which are the source of the characteristic odors of cruciferous plants, was tested first. Nine such oils were tested, and all attracted the larvae of the vegetable weevil. Five of six mustard oils were significantly attractive to adults. Volatile substances from umbelliferous plants, such as anethole, linalool, and limonene, were found to attract both larvae and adults. Apparently, chemical attractants aid polyphagous insects in locating host plants just as they do insects that are restricted to fewer hosts.

The onion maggot *(Hylemya antiqua)* is largely restricted to the onion, which, of course, produces many volatile, odorous sulfur compounds.[14] Two of those compounds, n-propyl disulfide and n-propyl mercaptan, were tested against adult females and found to be attractants. Field-trapping experiments demonstrated that they were powerful attractants for gravid females. Six sulfur compounds were tested against newly hatched larvae, and all were found to be attractants. One constituent of onion odor, methyl disulfide, was also a potent attractant for the black blowfly *(Phormia regina).*

Bark beetles are a very destructive group of forest insects, and thus much research has been done on the attractants produced by their hosts. The beetles are initially attracted to susceptible pine trees by volatile aldehydes or esters that are produced because of abnormal enzyme activity in subnormal trees. After isolated beetles make a few attacks on a suitable host, a second stronger attraction occurs, apparently because of the production of a pheromone by the beetles. Among the bark beetles members of only one sex select the host tree in which mating and subsequent oviposition will occur. In the genus *Dendroctonus* the entrance tunnel is excavated by the female, but in the genus *Ips* the male performs this function. In the case of *Ips confusus* in the ponderosa pine the secondary attractant was first localized in the frass material, which is a mixture of phloem fragments from the tree bark and excrement pellets of the beetle. Pheromone production occurs only after feeding commences, indicating that either a precursor is ingested and metabolized to the attractant or the metabolism of food material causes secretions in specialized cells. In any event, the secondary attractant causes a mass attack by the males on the host tree. Three terpene alcohols that were identified and synthesized are believed to be the principal components of the secondary attractant.[15] Mixtures of the three compounds are effective attractants in low concentrations, whereas each alone is not effective. This appears to be true for attraction of both males and females.

The principal pheromone in the frass produced by the female *Dendroctonus brevicomis* boring in ponderosa pine has also been identified and synthesized.[15] This compound, called exo-brevicomin, is active alone, though its activity is enhanced by a terpene, myrcene, and by other unidentified components.

J. A. Rudinsky supported the idea that many bark-beetle species recognize their host trees primarily through olfactory mechanisms responding to tree volatiles. Moreover, natural host-tree volatiles are utilized in the pheromone blends of each beetle species as synergists, or additives, that greatly increase the beetles' responses to the insect-produced components. Host terpenes are also a source of materials that are synthesized into the pheromones released by bark beetles, and several specific pathways have been identified.[16]

The skin of an apple has an attractant that causes larvae of the codling moth *(Laspeyresia pomonella)* to move toward the fruit. The attractant was identifed as an acyclic sesquiterpene, α-farnesene.[17,18]

Clayton Page and John Barber found that mosquito larvae *(Culex pipiens quinquefasciatus)* were attracted to all five of the plant species that they tested that had mucilaginous seeds, but were attracted to only half of the species with nonmucilaginous seeds.[19] In the latter group there was

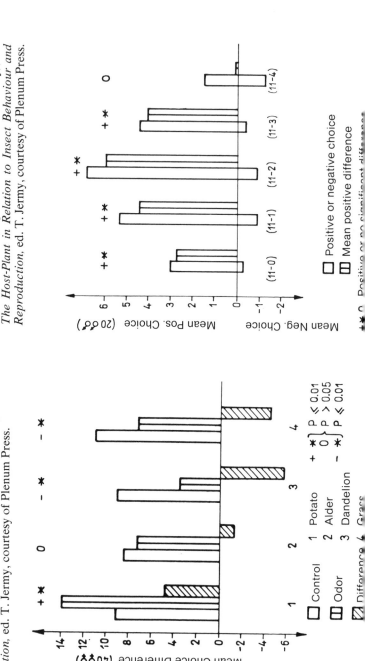

Fig. 7. Effect of nonhost-plant odor on the Colorado potato beetle, compared with the odor of potato foliage. Reproduced from J. de Wilde, "The olfactory component in host-plant selection in the adult Colorado beetle (*Leptinotarsa decemlineata* Say)," in *The Host-Plant in Relation to Insect Behaviour and Reproduction,* ed. T. Jermy, courtesy of Plenum Press.

Fig. 8. Role of antennal segments in anemotactic response in the Colorado potato beetle. 11-0, 11-1, 11-2, and so on, indicate 0, 1, 2, and more terminal antennal segments removed. Reproduced from J. de Wilde, "The olfactory component in host-plant selection in the adult Colorado beetle (*Leptinotarsa decemlineata* Say)," in *The Host-Plant in Relation to Insect Behaviour and Reproduction,* ed. T. Jermy, courtesy of Plenum Press.

considerable difference among varieties of the same species. In a wind tunnel walking Colorado potato beetles *(Leptinotarsa decemlineata)* show positive anemotactic responses (that is, they walk into the wind). Those responses are enhanced by potato-leaf odor, but reduced by the odor of several nonhost plants (fig. 7). Removal of the four terminal antennal segments eliminates the responses both to the wind and to the odors (fig. 8). Under field conditions walking Colorado beetles find potato plants from a distance of six meters when the plants are located upwind. J. de Wilde postulated therefore that the attraction of these beetles to their host plants in the potato family is probably caused by volatile chemicals.[20] No specific attractants were identified, however.

The odor of the host plant increases the activity of cabbage-root flies *(Erioschia brassicae),* but only gravid females move upwind toward the cabbage crop.[21] Gravid females showed the same response to allylisothiocyanate as they did to the cabbage plants, indicating that their upwind movement was caused by odor and not by a visual response to the cabbage plants.

The brassica pod midge *(Dasyneura brassicae)* is a serious pest on rape crops in Sweden. It causes important crop losses because suitable and efficient methods of control are lacking.[22] The midges fly passively with the wind until they are leeward of the host plants. Then they move upwind, usually at a low level. In a rape field the midges in the vegetation move upwind, but those in the air above the field move with the wind. Such behavior is to be expected when an olfactory stimulus is involved. Detailed experiments indicated that the females of this species are able to locate a host by odor, whereas the males do not react to the host-plant odor. The main organs for perceiving the odor are situated on the antennae.

Gy. Sáringer demonstrated conclusively that female

poppy beetles *(Ceutorrhynchus maculaalba)* are attracted to the garden poppy *(Papaver somniferum)* by the scent of the flowers. They are not attracted by the scent of the leaves or the shape or color of the flowers.[23]

The seeds of many plants contain appendages, known as elaiosomes, attached to the outside of the seed coats, and the outer layers of these appendages typically contain high concentrations of lipids that are often attractive to ants.[24] A diglyceride is apparently the attractant in the elaiosomes of the violet *Viola odorata.* Two diglycerides were identified —1,2-diolein and 1,3-diolein—and the ant *Aphaenogaster rudis* clearly preferred the former.

Much of the research on behavior-eliciting chemicals has stressed the role of one or, at most, a few chemicals in relation to a given activity. More recent studies suggest, however, that for many plant-feeding insects, such as lepidopterous larvae, host-plant recognition and acceptance are based upon complex sensory information that is both olfactory and gustatory. Even though a specific compound of a plant may contribute the chief sensory cue, it is the total chemical complex that is the basis for perception.[25]

Benzene extracts of the abdominal tips clipped from the virgin female gypsy moth *(Lymantria dispar)* strongly attract males of the same species. F. Acree isolated three active esters from the extract and reported that they are derived from at least two different alcohols.[26]

Social insects have extremely complex pheromone communication systems. The degree of social organization correlates with the frequency and complexity of the pheromones. Pheromones associated with the dynamics of a honey bee *(Apis mellifera)* colony include the following plus others: (1) queen pheromones that attract workers to the queen, where they are stimulated to feed and groom her; (2) attractant mating pheromones; (3) sting pheromones associated with nest defense; (4) foraging pheromones, apparently produced in the Nassanoff gland of the worker's abdomen; and (5) swarming and swarm-migration pheromones, which appear to be related to the production and distribution of queen pheromones.[27]

Honey bees, other bees, and many other social insects are often very important in agriculture, either as predators or pollinators or both, and chemical interactions in insect pollination will be discussed later in this chapter. Although much fascinating research has been done on insect pheromones, only those that most directly relate to the effects of the insects on crop plants will be discussed here. Readers interested in more details concerning the pheromones that are produced by other types of animals should refer to the works of Birch,[28] Méry,[29] Shorey,[30] and Müller-Schwarze and Mozell[31] in the Notes. Some details will be discussed in Chapters 6 and 7.

Arrestants

Dethier and his colleagues defined arrestants as chemicals that bring about aggregation in animals. Although such compounds generally have not been identified in host plants, the production of arrestant pheromones has been demonstrated in many insects. Obviously it is difficult to differentiate clearly between attractants and arrestants, and probably many compounds that are attractants become arrestants in higher concentrations.

When a piece of filter paper that had been used as a shelter in rearing a group of German cockroaches for several days was put into a jar with cockroach nymphs, the nymphs aggregated on this paper rather than clean filter paper. The aggregation occurred in darkness as well as light, but did not happen when the antennae of the nymphs were removed.[32] When feces collected from another batch of German cockroaches were placed on a piece of filter paper, the nymphs aggregated on that paper rather than a clean one. Moreover, a filter paper impregnated with an ether or methanol extract of German cockroach feces had the same effect. Feces collected from silkworm larvae had no such effect, however. The pheromone is apparently produced by the rectal pad cells of the cockroach, and it is lipid-soluble. It has not been identified.

As has been established, a single pheromone may serve as both an attractant and an arrestant. This is probably true of the pheromone produced by the initial bark-beetle invaders of conifer trees.[16] Wilde pointed out that the odor of potato plants appears to have an arrestant effect on flying Colorado beetles in addition to an attractant effect.[20] The sinigrin in rape plants appears to have a considerable arresting effect on flying brassica-pod midges, according to the research of J. Pettersson.[22]

Swarming in the honey bee is apparently related to the production and distribution of queen pheromones. Although swarms may leave the colony, move a short distance, and cluster temporarily, the pheromone complex of the queen is essential for normal clustering and swarm migration.[27] A swarm that leaves the colony without the queen soon returns to the colony. The so-called queen substance does not elicit swarming, but a related compound, 9-hydroxydec-*trans*-2-enoic acid, is very effective.

Stimulants

This topic has attracted much research activity in recent years because of the obvious practical significance in agriculture of stimulants to feeding, oviposition, and egg laying.

As early as 1905, A. Y. Grevillius found that larvae of the browntail moth *(Euproctis crysorrhoea),* which feeds on the chickweed *Stellaria,* could be induced to feed on other plants by smearing the leaves with a paste containing tannin, which is a constituent of chickweed.[33] Five years later Verschaffelt observed that the distribution of the mustard-oil glycoside among plants coincides with the range of acceptable host plants for larvae of the butterflies *Pieris rapae* and *P. brassicae*.[7] His experiments indicated that mustard-oil glycosides and their fission products stimulate feeding in these insects. Similar results were reported by Thorsteinson in his research with the diamondback moth *(Plutella maculipennis),* which is also specific to the members of the Cruciferae.[10] Adult vegetable weevils that do severe damage to crucifers were found to respond positively to five of six mustard oils in feeding tests.[14] Apparently, latex, or some substance contained therein, is the feeding stimulant that causes larvae of milkweed butterflies to eat milkweed plants.[9]

Eleven species of *Papilio* are known to feed on species of the Umbelliferae, a plant family characterized by the presence of essential oils, many of which are constituents of spices.[34] Studies indicated that at least some of these oils are feeding stimulants and also attractants for larvae of *Papilio ajax.*

Although the most efficient food plants of the tobacco hornworm *(Protoparce sexta)* are tobacco and tomato, this larva feeds widely and almost exclusively within the potato family.[35] A feeding stimulant, in the nature of a glucoside but not containing an alkaloid, has been isolated but not identified.

After a thorough study of the coevolution of butterflies and plants, Paul Ehrlich and Peter Raven concluded:

A systematic evaluation of the kinds of plants fed upon by the larvae of certain subgroups of butterflies leads unambiguously to the conclusion that secondary plant substances play the leading role in determining patterns of utilization. This seems true not only for butterflies but for all phytophagous [herbivorous] groups and also for those parasitic on plants. In this context, the irregular distribution in plants of such chemical compounds of unknown physiological function as alkaloids, quinones, essential oils (including terpenoids), glycosides (including cyanogenic substances and saponins), flavonoids, and even raphides (needlelike calcium oxalate crystals) is immediately explicable.[36]

The cucurbitacins are a class of bitter, toxic substances of almost universal occurrence in the cucumber family. These chemicals are classified as tetracyclic triterpenes, and selective breeding has eliminated them from the fruits of cultivated types of cucumbers, melons, and squashes. The first clue that the cucurbitacins might be feeding stimulants was the observation that spotted cucumber beetles *(Diabrotica undecimpunctata howardi)* fed voraciously upon sliced bitter fruits of a mutant watermelon strain *(Citrullus vulgaris)* obtained from the commercial variety Hawkesbury, whereas fruits of the original strain were much less preferred.[37] Controlled experiments with several of the fourteen known cucurbitacins indicated that all but one are feeding stimulants for the spotted cucumber beetle.

The beetle *Chrysolina brunsvicensis* is rather specific to its food plant, *Hypericum hirsutum.* It possesses a receptor that is especially sensitive to hypericin, which serves as the feeding stimulant in the plant.[38]

The superfamily Acridoidea comprises the locusts and short-horned grasshoppers, all of which are plant eaters.[39] These insects will feed in the complete absence of feeding stimulants, but only after being deprived of food for several hours. Under normal conditions biting and subsequent feeding occur only on specific host plants, and a feeding stimulant is required. Various kinds of nutrient chemicals, such as sugars, amino acids, and some fatty acids, are stimulants. Over 100 nonnutrient chemicals were tested on *Locusta* and *Schistocera.* None enhanced feeding by *Locusta,* but several chemicals, belonging to different chemical classes, enhanced feeding by *Schistocera.* In tests with *Schistocera* two alkaloids, hordenine and lobeline hydrochloride, served as feeding stimulants in low concentrations, but were deterrents at higher concentrations. A sapogenin, diosgenin, was stimulatory at all concentrations tested.

Since the onion maggot is largely restricted to the onion, females of this species always select onion plants for oviposition. Therefore some organic sulfur compounds occurring in onion plants were tested for oviposition by the maggots. N-propyl mercaptan and n-propyl disulfide significantly increased the numbers of eggs laid in dishes containing sand. Of four alcohols and five carbonyl compounds tested, all of which are recorded as constituents of onion odor, only n-propyl alcohol and acetone slightly stimulated egg laying.[14]

In laboratory bioassays gravid female codling moths *(Laspeyresia pomonella)* were stimulated to oviposit by natural α-farnesene and synthetic (E,E)-α-farnesene. α-Farnese is composed of (E,E)-α- and (Z,E)-α-farnesene.[40]

The beet moth *(Scrobipalpa ocellatella)* is restricted in its host plants. The larvae feed on plants belonging to the genus *Beta* and especially on cultivated beets *(B. vulgaris).* The females lay eggs on these plants also. The beet plant, or an aqueous extract of beet leaves, was found to stimulate oogenesis, release oviposition, and determine the choice of the egg-laying site.[41] An aqueous leaf-extract of the chestnut tree *Castanea sativa,* when sprayed on a sugar-beet plant, masks the stimulating aura of the beet. Then the beet plant will no longer stimulate oogenesis or release oviposition but acts instead like a nonhost plant. Only extracts of the chestnut leaves have such an effect; extracts of the fruit or wood have no effect at all.

Females of the poppy beetle lay their eggs after the poppy petals fall, when the young seed capsules become accessible. First they peel away the outer wall of the capsule, then they gnaw a hole in the peeled surface and lay their eggs on the septum walls inside the capsule. Oviposition is

77

not performed in capsules that are more than three days old.[23] The gravid females are attracted to the flowers by chemicals produced by the petals, and apparently substances present in the poppy capsules stimulate the females to feed on the capsules and to oviposit.

Fecundity, in terms of the number of eggs laid, is increased in the Indian meal moth *(Plodia interpunctella)* by the odor of certain chemicals.[42] If 50 percent of the moth's antennae segments are removed, however, the olfactory stimuli lose their effect, and fecundity is the same as if there were no stimulation. This insect is harmful to many stored products. Eighty-three different kinds of food have been found to be eaten by meal-moth larvae.

In the absence of beans virgin females of the bean weevil *(Acanthoscelides obtectus)* do not lay eggs. The presence of the host plant induces oviposition and also stimulates oogenesis.[43]

The European elm bark beetle *(Scolytus multistriatus)* is the principal vector of Dutch elm disease. Its most potent feeding stimulant is hydroquinone.[44]

Many other examples of stimulants could be cited from the voluminous literature on plant-insect and insect-insect chemical interactions, but those discussed above should suffice to demonstrate the widespread distribution and importance of such compounds.

Repellents

Populations of herbivorous insects and the plants upon which they feed are subject to many destructive forces, ranging from unfavorable climatic and soil conditions to attack by pathogens, parasites, and predators.[45] There is good reason to believe that plants have survived largely because of defensive mechanisms that they have evolved through time in relation to the attacks of herbivores and pathogens. As has been shown above, there are three basic reasons to suspect that coevolution between insects and plants has a substantial biochemical component based on nonnutrient compounds:

(1) Many insect species discriminate between host and nonhost plants largely through differential responses to various secondary plant compounds.

(2) Many of the secondary substances in plants are known to be poisonous to insects and other animals.

(3) The various food plants of a particular insect species, genus, or even family often share similar secondary compounds, even though they may differ greatly in other respects.[45]

Repellent and deterrent chemicals are obviously of great importance in agriculture because they protect crop plants against parasites and predators. Most breeding programs concerned with protection of crop plants against insects have thus concentrated on these chemicals. Unfortunately, it is sometimes very difficult to determine whether a given chemical constituent of a host plant is a true repellent that will cause an insect to make oriented movements away from the source, or whether it is just a deterrent against feeding, oogenesis, or oviposition. Many of the defense pheromones secreted by insects have been clearly identified as repellents, but these have less direct impact in agriculture. They include such substances as benzoquinones, aromatic compounds, terpenoids, acids, phenols, esters, carbonyl compounds, steroids, and others.[46]

Three species of flour beetle, *Tribolium confusum, T. castaneum* and *T. destructor,* and the long-headed flour beetle, *Latheticus oryzae,* produce relatively large amounts of 2-ethyl-1,4-benzoquinone and usually one or two additional quinones. The ethylquinone was found to be a potent repellent of these beetles.[47,48] In 1966, Martin Jacobson listed thirty-seven repellent compounds known to be produced by fifty-eight species of arthropods, of which all but sixteen were insects.[49]

The chemical compound DIMBOA (2,4-dihydroxy-7-methoxy-1,4-benzoxazine-3-one) was demonstrated in 1967 to be a factor in the resistance of corn to the European corn borer.[50] Subsequently five inbreds of dent corn were analyzed at various stages of plant development for concentrations of DIMBOA.[51] Concentrations were high in the embryonic plant, but decreased as the plant matured. The concentrations were generally higher in the root than they were, in decreasing order, in the stalk, whorl, and leaf. The high concentration of DIMBOA in seedling corn apparently explains the resistance of young corn plants to the European corn borer. Those inbreds that maintained high concentrations of the compound at later stages of development were borer-resistant. Unfortunately it is not clear whether this compound acts as a repellent or a deterrent, or both.

Glandular trichomes, or hairs, are very common throughout the dicotyledonous angiosperms, and they secrete a variety of secondary plant products, such as terpenoids, tannins, phenolics, quinones, alkaloids, and flavonoids.[52] There is direct evidence that some volatile components of trichome exudates serve as repellents to certain insects. For example, oil of citronella, an essential oil in the grass *Andropogon nardus,* has been employed as a mosquito repellent for many years. It is composed primarily of geraniol, with citronellol, citronellal, and borneol among its minor constituents. Oil of eucalyptus, thyme, sassafras, pennyroyal, clove, wintergreen and anise are other insect repellents that occur in certain plants.

A. R. Penfold and F. R. Morrison tested forty essential oils from numerous plant species in Australia as repellents against mosquitoes, March flies, and sand flies.[53] The most effective oils were the oil of Huon pine wood *(Dacrydium franklini)* and leaf oils from *Backhousia myrtifolia, Melaleuca bracteata,* and *Zieria smithii.*

The terpene α-pinene from pine is strongly repellent to

the pine bark beetle *Hylurgops palliatus*. It is weakly repellent in relatively high concentrations to another bark beetle, *Hylastes aster*.[49]

Cotton seedlings and squares were found by F. G. Maxwell and coworkers to contain volatile materials that are repellent to the adult boll weevil, *Anthonomus grandis*.[54] Cotton seedlings that were painted with an emulsified concentrate of the materials effectively repelled all weevils for five hours, and only medium damage was done after twelve hours. Repellency appears to be associated with the pungent odor of the material, since physical contact is not necessary. The repellent is stable to heat and is not toxic to the cotton seedlings.

T. Eisner reported that seventeen species of insects were repelled by the vapors of nepetalactone, a compound isolated from the catnip plant *(Nepeta cataria)* that is related chemically to several lactones isolated from insects.[55] Menthol, which emanates from the trichomes of many plants of the mint family, is strongly repellent to the silkworm.[52]

T. H. Hsiao concluded that the selection of food plants by the larvae of the Colorado potato beetle is influenced mainly by qualitative and quantitative differences in the deterrent and repellent chemicals of the food plants (fig. 9).[56]

Experiments in the Soviet Union demonstrated that the damage done by the bark beetle *Callidium violaceum* in unbarked coniferous lumber stored for one year averaged 95 percent of the volume of the spruce lumber, 60 percent of the pine, and 30 percent of the fir.[57] There was seven to eight times as much volatile oil in fir phloem as in spruce, and the toxicity of the oil was 3.2 times higher in undamaged fir than in spruce. In decreasing order the following substances were toxic to larvae in the wood: bornil acetate, $\triangle$3-carene, α-pinene, ß-pinene, and limonene. The resistance of fir lumber to this bark beetle was found to increase with an increase in bornil acetate, the most toxic terpene. Ob-

Food Plant

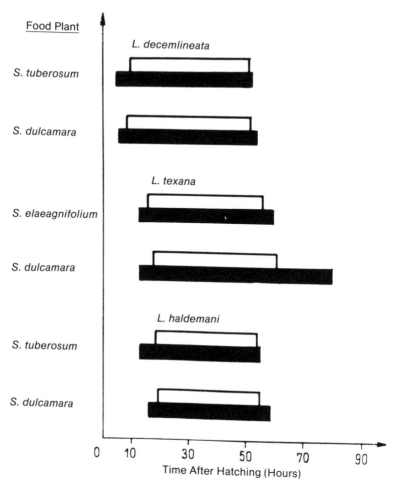

Fig. 9. The influence of feeding habits on the rate of growth of first-instar larvae of *Leptinotarsa* species fed on different *Solanum* (potato-genus) plants. The data represent the averages from rearings of 50 larvae on each plant. The upper and lower bars represent, respectively, the larvae that have and have not consumed their own egg shells before feeding on the plant. Each bar shows the time of initiation of feeding on the food plant and the time of molting into the second instar. Reproduced from T. H. Hsiao, "Chemical and behavioral factors influencing food selection of *Leptinotarsa* beetle," in *The Host-Plant in Relation to Insect Behaviour and Reproduction,* ed. T. Jermy, courtesy of Plenum Press.

viously, it is difficult to infer whether the bornil acetate acts primarily as a repellent, a feeding deterrent, or an insecticide. Perhaps the same compound fits each of those categories. Pyrethrum is a plant substance that is both repellent and insecticidal.[46]

Dale Norris implemented a very complex project to determine what chemicals deter the elm bark beetle *Scolytus multistriatus* from alighting and feeding on nonhost trees.[58] Thirteen species in nine families and ten genera were analyzed. Even though the complete complement of allelochemicals in the various species of trees was complex and variable, there were similar structural and/or electrochemical molecular characteristics among the identified chemicals. Norris's initial studies of hickories *(Carya)* indicated that juglone (5-hydroxy-1,4-naphthoquinone) keeps *S. multistriatus* off the bitternut hickory *(C. cordiformis)* and shagbark hickory *(C. ovata)*. This compound occurs mostly as a glucoside in intact healthy cells of hickories, though it occurs in perceptible amounts in the atmosphere surrounding healthy trees. Additional amounts of juglone are released into the atmosphere if cells are ruptured, if the plant becomes stressed (as during drought), or if the plant becomes diseased. According to Norris, once a hickory tree becomes irreversibly diseased, it no longer can release juglone, and secondary predators and parasites that attack dying trees of several species appear on the plant.

Subsequent studies by Norris on the black walnut *(Juglans nigra)* demonstrated that juglone is the main repellent against *S. multistriatus* in this tree species also. Several 1,4-naphthoquinones were tested against the elm bark beetle, and the repelling effect was found to increase with an increase in the oxidation-reduction potential of the compound. Hydroxyl (OH) substitutions always made the naphthoquinone more repellent, or inhibitory to feeding, than the relative redox potential would indicate.

An apparent boll-weevil repellent was found in an attractant extract from the cotton plant. Removal of the repellent improved the effectiveness of the attractant, and initial work suggested that an attractant-repellent ratio may exist in cotton and other hosts of the boll weevil that helps determine whether a particular host is selected by the weevil.[59]

In general, it appears that only those compounds that are relatively volatile serve as true repellents in plant-insect chemical interactions. Volatility is not a requirement of many insect-produced repellents because these are often forcibly ejected toward the potential enemy.

Deterrents

Many glycoalkaloids occur in various species of the potato family, and various of them are feeding deterrents or even toxic to the Colorado potato beetle. Solanine and chaconine occur in the potato and have no apparent effect on the potato beetle.[35] Tomatine in tomato, the leptines in *Solanum chacoense,* soladulcin in *S. dulcamare,* and a tetrosid in *S. acaulia* are feeding deterrents to the potato beetle. Two other compounds occur in some species of the potato family and serve as feeding deterrents and sometimes as toxins to the potato beetle. They are capsaicin, which is the pungent principle in red peppers, and nicotine in tobacco. Because nicotine is synthesized in the roots of tobacco plants, a tobacco plant grafted on the root of a potato plant is free of nicotine and will be eaten by the potato beetle. On the other hand, a potato plant grafted on a tobacco root becomes resistant to the Colorado potato beetle.

One natural feeding deterrent was identified as a mustard oil (2-phenylethylisothiocyanate) in the turnip root *(Brassica napus)* and as myristicin (5-allyl-1-methoxy-2,3-methylene-dioxybenzene) in the parsnip root *(Pastinaca sativa).*[60,61]

These compounds were shown to be deterrent and, indeed, insecticidal to vinegar flies *(Drosophila melanogastor)*, house flies *(Musca domestica)*, flour beetles, mosquito larvae *(Aedes aegypti)*, spider mites *(Tetranychus atlanticus)*, pea aphids *(Acrythosiphon pisum)*, and Mexican bean beetles *(Epilachna varivestis)*.

Larvae of the moth, *Euchelia jacobaeae* feed on many species of groundsel *(Senecio)*, but not on *S. viscosus,* which is densely covered with glandular hairs.[62] When the glandular secretion is dissolved away with alcohol, however, the larvae feed on this species. When the same substance is spread on the leaves of a usually acceptable species of groundsel, these plants are rejected as food plants.

The rose of sharon *(Hibiscus syriacus)* has a strong feeding and oviposition deterrent against the cotton boll weevil.[63] When the calyx of the flower is removed, however, the boll weevil feeds and oviposits as well on this plant as on cotton. From an extract of the calyxes a nonvolatile water-soluble material was isolated that strongly deterred feeding by the weevils. A substance that deters feeding of the boll weevil was also found in the filtrate of a water extract of tung meal, which is a byproduct of extracting oil from the seeds of the tung tree *(Aleurites fordii).*[64] The same substance, or a related one, is present in a more concentrated form in tung oil. The deterrent is readily soluble in water, resistant to moderate heat, and volatilizes slowly. The increase in deterrent effect is directly proportional to an increase in concentration of the substance.

The coral bead *Cocculus trilobus* is well known as the host of two Japanese fruit-piercing moths but is not attacked by any other insects in nature.[65] Two alkaloids were isolated from fresh leaves of the plant. One was identified as the feeding deterrent isoboldine, and the other as an insecticide named cocculolidine.

It was noted in Japan that the leaves of the shrub kusagi

(Clerodendron tricotomum) are not eaten by insects. Feeding tests with selected insects confirmed that the leaves contained feeding deterrents, and two compounds were isolated, named clerodendrin A and clerodendrin B.[65] Relatively small amounts, 200 to 300 parts per million, deter the feeding of insects on their usual hosts. These deterrents may prevent insect attacks on rice plants.

Leaves of the Japanese shrub shiromoji *(Parabenzoin trilobum)* are not attacked by larvae of the insect *Prodenia litura,* and a crude extract prevents feeding on the leaves of that insect's usual hosts.[65] Two active feeding deterrents were isolated from the extract, named shiromodiol-diacetate and shiromodial-monoacetate.

The cucurbitacins, discussed above as potent feeding stimulants for the cucumber beetle, are also feeding deterrents for the honey bee and the yellow jacket.[37] The Boston fern leaf contains a water-soluble feeding deterrent effective against the southern armyworm.[66]

The possible relationship between the concentration of DIMBOA in corn and its resistance to the European corn borer was discussed above among the repellents. DIMBOA occurs in the corn plant, however, only in combination with glucose, that is, as a glucoside. When the plant tissue is crushed by the mandibles of the borer, the glucoside is rapidly digested to glucose and DIMBOA. Therefore, it may be that DIMBOA serves mostly as a feeding deterrent and only slightly as a repellent. It also has an insecticidal property. European corn borers that were fed resistant corn tissue developed more slowly, weighed less, mated less successfully, and produced fewer egg masses than those fed susceptible tissue.[67] DIMBOA also has been implicated in the resistance of corn to the corn earworm, *(Heliothis zea).*[63]

Even though cholesterol is known to be essential in the diet of many insects, G. M. Chippendale and G. P. V. Reddy found that cholesterol acetate, cholesterol myristate, and cholesterol oleate are feeding deterrents to the southwestern corn borer *(Diatraea grandiosella).* Cholesterol had no effect.[68]

In populations of bird's-foot trefoil *(Lotus corniculatus)* some individuals contain a cyanogenic glucoside that produces hydrogen cyanide when the plants are injured. These individuals are bothered by fewer insects than their noncyanogenic neighbors.[69]

Locusts and grasshoppers not previously deprived of food may reject some potential food plants without biting them, after their maxillary palps, or mouthparts, have simply touched the plant surface. *Chorthippus parallelus* rejects certain plants following contact of the leaf surface by the maxillary palps, and a similar response was demonstrated in the case of *Chortoicetes terminifera.*[70, 71] Experiments with extracts of leaf-surface materials from annual bluegrass *(Poa annua)* and English daisy *(Bellis perennis)* showed that locusts *(Locusta)* can distinguish between the waxes of different plants.[72] The insect *Acyrthosiphon pisum* responds to certain chemicals (n-alkanes) from the surface of the broad bean *(Vicia faba)* just by probing and can differentiate between related compounds with different chain lengths.[73]

Several calabash trees *(Crescentia alata)* were defoliated by hand during the rainy season in Costa Rica. The new leaves produced by these trees were severely attacked and eaten by adult flea beetles (*Oedionychus* sp.), though the leaves on control trees were not eaten.[74] Some of the trees were completely defoliated by the beetles. All of the experimental trees put out new leaves and again were attacked by the beetles, though not as severely as the first time. Flea beetles are host-specific, and apparently the host leaves of this beetle are edible for only a short time, when they are immature. It was suggested that increases in certain deterrent secondary plant products may make the leaves inedible later.

In Britain more species of butterflies and moths have been recorded feeding on the leaves of the common oak *(Quercus robur)* than on the leaves of any other tree species.[75] The majority of the species are found on oak leaves in the spring, and high densities are frequently attained in May. The infestations are occasionally on such a scale as to cause complete defoliation of oak trees in late May and early June. The concentration of tannins in the leaves was found to increase from 0.5 percent of leaf dry weight in April to about 5.0 percent in September. Condensed tannin did not appear in the leaves until late May.[76] Thus the period of highest insect attack on the oak leaves occurs when the total concentration of tannin is low and when condensed tannin is virtually absent.

The effect of the age of the plant or plant structure on the concentration of feeding deterrents is often the reverse of that described for calabash and common oak trees. Grass seedlings are relatively distasteful to locusts and grasshoppers, compared with mature leaves of the same species.[77] After five hours without food the meals of *Locusta migratoria* nymphs are up to nine times larger when the locusts feed on mature leaves than when they eat seedling leaves of the same species. This phenomenon was tested over a range of twenty grasses (fig. 10). It is usually four weeks after sowing before young plants are eaten in the amounts that are normal for mature grasses. Halostachine, an alkaloid that deters feeding by *Locusta,* is present in high concentrations in young seedlings of rye grass *(Lolium perenne),* but its concentration decreases as the plant matures. The alkaloids, gramine and hordenine, that occur in species of barley (*Hordeum* spp.) are also found in highest concentrations in the leaves of seedlings. Both of those chemicals are feeding deterrents for the grasshopper *Melanoplus bivittatus.* It should be recalled that DIMBOA, the feeding deterrent against the Eu-

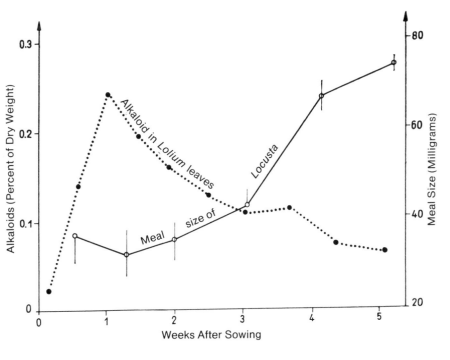

Fig. 10. The relationship of the meal size taken by *Locusta* nymphs and the alkaloid content of the perennial ryegrass *(Lolium perenne).* Reproduced from E. A. Bernays and R. F. Chapman, "Antifeedant properties of seedling grasses," in *The Host-Plant in Relation to Insect Behaviour and Reproduction,* ed. T. Jermy, courtesy of Plenum Press.

ropean corn borer, also occurs in higher concentrations in corn seedlings than in older plants.

In the midwestern United States bacterial wilt caused by *Erwinia tracheiphila* is transmitted to cantaloupe *(Cucumis mello)* primarily by the striped cucumber beetle *(Acalymma vittata)*. Therefore 1,643 introductions of *Cucumis mello,* 19 introductions of other *Cucumis* species, and 62 American varieties of commercial cantaloupe were screened for feeding deterrents, or resistance to that beetle.[78] No feeding or only light feeding occurred on 13.3 percent of the *C. mello* introductions, 42 percent of the other *Cucumis* species, and 29 percent of the commercial varieties. The feeding deterrents were not identified.

Several flavonoids from apple trees, cottonwood *(Populus deltoides),* black locust *(Robinia pseudoacacia)* and bur oak *(Quercus macrocarpa)* were found by Norris to deter feeding by the elm bark beetle.[58] Coumarins from horse chestnut *(Aesculus octandra)* and white ash *(Fraxinus americana)* deterred feeding and may perhaps be repellents. In addition, two alkaloids, gramine from box elder *(Acer negundo)* and silver maple *(A. saccharinum)* and magnoline from *Magnolia acuminata,* are feeding deterrents of the elm bark beetle.

Relatively large amounts of benzyl alcohol occur in the strains of barley that are resistant to the greenbug *(Schizaphis graminum),* but there is none, or very little, in susceptible strains.[79] If flats containing susceptible strains of barley are watered with a 0.01 percent solution of benzyl alcohol before the addition of the greenbugs to the plants, no damage occurs. Thus it appears that this compound is a feeding deterrent to greenbugs.

Certain compounds that foam in water solutions and are bitter to the taste are known as saponins. Saponins from legume seeds have been demonstrated to be feeding deterrents to the aphid *Myzus persicae.*[80] It seems possible therefore that saponins may be important deterrents to some seed-eating insects in addition to being toxic to them. The naturally occurring insecticidal saponins are discussed further in the first part of Chapter 6.

Sesquiterpene lactones are terpenoid compounds with fifteen carbon atoms, commonly referred to as "bitter principles." This suggests that they may act as feeding deterrents. There are over 600 known sesquiterpene lactones, of which about 90 percent are from the composite (Compositae) family of plants. One of the compounds, glaucolide-A, occurs in certain species of ironweed *(Vernonia)*. When the weed was fed to larvae of six butterfly and moth species in tests, the compound was a feeding deterrent to all, though its strength varied somewhat among the species.[81] The compound was also found to deter oviposition in gravid females of several other moth species.

The locust *Locusta migratoria* feeds on most grasses but almost invariably rejects nongrasses without feeding. All of the nonhost plants yielded extracts that deterred feeding when added to wheat-flour wafers, which are intensely stimulatory to this locust. The grasshopper *Chorthippus parallelus* also feeds on grasses, and its responses were similar to the locust's in tests of extracts from nonhost plants. The highly palatable grass *Holcus* is refused by this grasshopper if the grass is allowed to take up coumarin from a water solution. E. A. Bernays and R. F. Chapman concluded that the presence or absence of feeding deterrents determines whether a plant is eaten or not by the grass-eating grasshoppers and locusts.[39] If no deterrents are present, feeding is initiated and maintained by feeding stimulants and continues until it is stopped by feedback from the foregut stretch receptors.

A fascinating extension of the protection of plants by chemical compounds is the accumulation in certain insects of plant-host-produced compounds that make those insects resistant to certain predators. A classic North American example is the monarch butterfly *(Danaus plexippus)*, whose coloring warns potential predators that it is unpalatable.[82] Experimental evidence indicates that several species of birds, in at least two families, find the monarch unacceptable, compared with controls. The butterfly feeds on plants of the milkweed (Asclepiadaceae) family, which is well known pharmacologically because many species contain large quantities of cardiac glycosides, which are potent heart poisons. The general effect of the compounds on vertebrates is to decrease the frequency of the heartbeat and increase its amplitude. A large dose given intravenously will cause heart failure and death, but, if the dose is given orally, nausea and vomiting prevent death. Monarch butterflies that are fed exclusively on such a milkweek contain large quantities of these heart poisons,[82] and L. P. Brower and his colleagues have demonstrated that the compounds are accumulated from the host and not made in the insect. Certain bird predators learn to reject monarchs after one or more emetic experiences. Researchers have reported a few other examples of such phenomena.[83]

Chemicals with Delayed Effects

Many chemical compounds that insects assimilate from plant hosts or other food sources do not elicit immediate behavioral responses, but have very important delayed effects on growth, hormonal changes, reproductive differentiation, and other life functions. Some compounds have both an immediate and a delayed effect on the same species of insect, and one compound will affect different insects differently.

As described above, butterfly and moth larvae feed on oak leaves chiefly in the spring, apparently because the total tannin content of the leaves is lower in the spring, and condensed tannin is virtually absent during that season. Winter-moth larvae *(Operophtera brumata)* were grown in the laboratory on an artificial diet containing casein.[75] The addition to the diet of as little as 1 percent tannin from September oak leaves caused a significant reduction in larval growth rate and pupal weight. It was not possible to determine whether the tannin ultimately reduced adult weight and fecundity, because of high mortality rates during the experiments. There was evidence, however, that pupal survival was reduced.

The greenbug is one of the most damaging insects to small grains in the United States. Under favorable environmental conditions it can severely damage or destroy susceptible crops. An understanding of the nature of greenbug resistance in host plants is important in selective breeding for control of this pest. As described above, benzyl alcohol is a feeding deterrent in certain resistant varieties of barley. Cereals produce a great many secondary plant substances, of which many have been demonstrated to increase insect resistance. For example, Glenn Todd and his coworkers tested a large number of flavonoids and some derivatives of benzene, benzoic acid, benzaldehyde, and cinnamic acid, plus some miscellaneous compounds, for their effects on greenbug growth and reproduction.[84] Very dilute amounts of the test compounds (3.75×10^{-4} M or less) were incorporated in a chemically defined diet. Many of the compounds reduced greenbug growth, numbers of progeny, and survival of the progeny. Survival was less than 20 percent among the greenbugs feeding on diets that included vanillic, sinapic, syringic, gentisic, or ferulic acids. Because many of those compounds occur in barley leaves, it appears likely that at least some of the resistance of barley to greenbugs may be owing to the presence of some of the phenolic and flavonoid compounds.

The larval food range of the black swallowtail butterfly is restricted to plants of the carrot family. Of course, plants of many other families, including the Cruciferae, also occur in the gardens and natural habitats frequented by this butterfly. To determine why black swallowtail larvae do not eat species of the mustard family under natural conditions, J. M. Erickson and P. Feeny reared larvae on celery leaves that were cultured in solutions of sinigrin, a common mustard-oil glucoside in the mustard family.[85] Feeding rates were not appreciably affected, but growth and development were markedly reduced. At a concentration of 0.1 percent or more of the fresh-leaf weight sinigrin caused 100 percent larval mortality. Thus the mustard-oil glucosides appear to have a defensive function in the mustard family.

J. C. Bailey and his colleagues implanted boll-weevil eggs in buds of twelve lines of cotton representing five species and one interspecific cross.[86] They found that significant differences existed in several lines in the percent of adult emergence, adult weight, and the number of days required for adult emergence. Obviously the lines that were wholly or partially resistant had characteristics that caused delayed effects on the boll weevils, though the characteristics were not identified in the tests.

The leafhopper, or jassid *(Amrasca devastans)*, is one of the most serious pests of cotton in India. It also feeds on other crop plants, including okra *(Hibiscus esculentus)* and castor bean *(Ricinus communis)*. A study was undertaken to determine the mechanism of resistance of certain varieties of cotton to this insect.[87] Of the six varieties of cotton studied during the peak period of jassid activity, two were highly resistant, two had medium resistance, and two were highly susceptible. It was demonstrated that these insects preferred to oviposit on the susceptible varieties instead of on the resistant varieties. Moreover, the survival rate of nymphs was 33 to 50 percent on the highly resistant varieties, 75 percent on the varieties with medium resistance, and 92 to 96 percent on susceptible varieties. The nymphal period was also 3 to 4 days longer on resistant varieties than on susceptible ones. No concrete evidence was given to explain the delayed effects of resistant varieties. It is very possible that certain secondary plant substances may have been responsible, and this should be investigated.

In all species of the cotton genus *Gossypium* glands occur on the plant parts that are abovegound. The contents of these glands are known to be toxic to nonruminant animals, including insects.[6] Three chemical compounds from the glands—gossypol, quercetin, and rutin—were found by M. J. Lukefahr and D. F. Martin to inhibit the larval growth of the cotton bollworm and the tobacco budworm, and all three markedly decreased the numbers of larvae reaching the pupal stage.[88] Less than 30 percent of the larvae of both species reached the pupal stage when the diet contained 0.2 percent gossypol. Quercetin was even more toxic to the tobacco budworm; less than 30 percent of the larvae reached the pupal stage if their diets contained as little as 0.1 percent of this compound. Quercetin was much less toxic to the bollworm, however, than gossypol. Rutin was less toxic to both species than quercetin or gossypol. These researchers suggested that cotton breeders might select cotton plants with high amounts of these compounds to increase resistance to bollworms. High-gossypol lines have been found among the dooryard cottons.[6]

J. C. Reese and S. D. Beck[89, 90] investigated the chronic effects on the black cutworm *(Agrotis ipsilon)* of small amounts of various allelochemics. This insect was chosen for study because it feeds on a wide variety of host plants and is thus exposed to a wide variety of secondary plant compounds. By incorporating them into a synthetic diet, Reese and Beck examined the effects of thirty-seven common compounds, most of which were common phenolic compounds in plants, or their oxidized forms (quinones). Of the thirty-seven only ten compounds were toxic, while twenty-five reduced growth, pupation, or pupal weight. In other words, they had delayed or chronic effects. Some of the chronic effects did not save the host plant from being partially consumed, but, where there are enough deleterious effects on the insect species, there may not be large populations attacking the particular plant. According to Reese and Beck, such a partial resistance is common, even among host plants, because highly susceptible plants do not survive long, and host plants usually contain at least small amounts of various deleterious chemicals.[89]

A flour-beetle larval-protease inhibitor has been isolated from soybean, and such inhibitors are known to be present in legume species that are generally regarded as resistant to damage by flour beetles (a protease is a protein-digesting enzyme).[6] Because these inhibitors are specific for a given flour-beetle species, it should be possible to breed seed varieties containing higher concentrations of the specific protease inhibitors required. Thus they will afford at least some resistance to damage in storage.

Five different saponins from soybean seeds inhibited larval growth in the flour beetle *Tribolium costaneum* and prevented pupation of the larvae of the cowpea seed beetle *Callosobruchus maculatus.*[80] Eleven alkaloids were tested against the larvae of *C. maculatus* by adding them to the beetle's regular cowpea seed-meal diet. Nine were found to be totally lethal at the lowest concentration tested, 0.1 percent; and the other two markedly reduced beetle production at the same low concentration.[91] Of twenty-four nonprotein amino acids found in seeds, most were lethal to the cowpea-weevil larvae at a 5 percent concentration in the food. Several significantly reduced the number of emerging adult beetles.

The effects of ingestion of a sesquiterpene lactone, glaucolide-A, on the growth and development of several moth and butterfly larvae were tested by adding the compound to an artificial diet in concentrations up to 0.5 percent of the dry weight (fig. 11).[81] The larvae tested were the southern

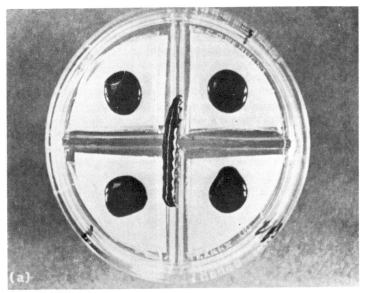

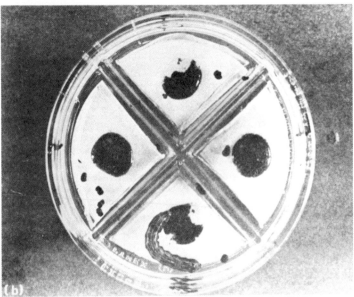

Fig. 11. Larval-feeding preference test of southern armyworm *(Spodoptera eridania)* on *Vernonia* powder-agar discs made from different
species of that genus: (a) initiation of test, (b) completed test. Reproduced from Burnett et al., "The role of sesquiterpene lactones
in plant-animal coevolution," in *Biochemical Aspects of Plant and
Animal Coevolution,* ed. J. B. Harborne. Copyright: Academic Press
Inc. (London) Ltd.

armyworm *(Spodoptera eridania),* the fall armyworm *(S. fru-giperda),* the yellow-striped armyworm *(S. ornithogalli),* the cabbage looper *(Trichoplusia ni),* and the yellow woollybear *(Diacrisia virginica).* After fourteen days each surviving larva was weighed and then allowed to complete its life cycle on the same diet. The growth of the southern, fall, and yellow-striped armyworms was greatly reduced when restricted to diets containing glaucolide-A. These same species avoided diets containing that compound in larval feeding tests. The cabbage looper and the yellow woollybear were not reduced in weight on diets containing glaucolide-A (fig. 12). With this compound in the diet the number of days to pupation was increased for all species except the yellow woollybear. Thus some insects feeding on plants containing sesquiterpene lactones would have their life cycles extended, increasing their exposure to predators, parasites, pathogenic microorganisms, and adverse meteorological conditions. They would also have fewer generations during the active season, thus lowering the sizes of their populations. The southern and fall armyworms showed sizable reductions in the percentage of surviving larvae, pupae, and adults when reared on diets with glauco-lide-A. The reduction in survivors was especially evident during the early instars.

There has obviously been much less research done on the delayed effects than on the immediate behavioral effects of secondary plant products on insects. Some of the remarkable results indicate the pressing need for such research.

CHEMICAL ATTRACTANTS AND INSECT POLLINATION

Plant species that are of economic importance are of two kinds: either they are self-fertile and set fruit or seed through self-pollination, or they are self-infertile and must be cross-pollinated by other plants of the same species in order to

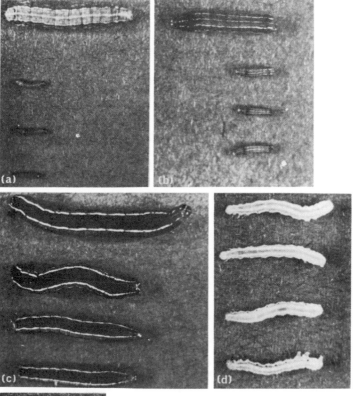

Fig. 12. Relative sizes of larvae reared on diets containing glaucolide-A. Top insect in each photograph fed on control diet followed by increasing glaucolide-A concentrations of 0.125%, 0.25%, and 0.5% by dry weight. (a) fall armyworm; (b) southern armyworm; (c) yellow-striped armyworm; (d) cabbage looper; (e) yellow woollybear. Reproduced from Burnett et al., "The role of sesquiterpene lactones in plant-animal coevolution," in *Biochemical Aspects of Plant and Animal Coevolution,* ed. J. B. Harborne. Copyright: Academic Press Inc. (London) Ltd.

produce seeds and fruits. Some self-fertile species are automatically pollinated with pollen from their own flowers. Others are so constructed that assistance from wind or insects is required to transfer the pollen from their anthers to their stigmas. Moreover, self-fertile plants may produce more or better-quality seeds and fruits if they are cross-pollinated rather than self-pollinated. Wind is the principal pollinating agent for agricultural grasses and a few other grass species, whereas most crop plants with conspicuous, colored, and scented flowers are insect-pollinated.[92] Thus the attraction of appropriate pollinator insects by flowers is very important in agriculture.

The most important pollinating insects in agriculture are solitary bees, bumblebees, and honeybees.[92] Other insects often visit the flowers of crop plants, but most are not believed to be important in pollination generally. They do not have sufficient body hair and the necessary behavior patterns to transfer pollen from the anthers to the stigmas of the flowers that they visit. Also, unlike the bees, which forage consistently to obtain food for their young, most other insects forage only to satisfy their own immediate needs and use a variety of foods other than those from flowers. It is thought that many of the more important supplementary pollinators are two-winged flies (Diptera), including various species of the genera *Eristalis, Syrphus, Platycheirus, Rhingia, Calliphora, Lucilia, Sarcophaga, Bibio, Dilophus,* and *Bombylius.*[92]

Under natural conditions pollinating insect populations usually are sufficiently large to pollinate the native plants. When large fields are occupied by a single flowering crop, however, there may be too few wild pollinators, and the yield may be limited by lack of pollination. Moreover, clean and intensive cultivation has eliminated many of the natural food sources and nesting sites of wild pollinating insects. Widespread use of pesticides may destroy many natural pollinators also. For most crops a deficiency of wild pollinators can only be supplemented by honeybee colonies. Honeybees have been kept in hives since time immemorial because of their production of honey and wax. Thus they can be made readily available for pollination when needed. It has been estimated that they are several times more valuable as pollinators than as producers of honey and wax.[92] The pollinating potential of a single colony is great because its bees will make up to four million trips from the hive each year and visit an average of one hundred flowers on each trip.

As previously stated, F. Huber demonstrated as early as 1821 that bees are attracted to flowers by their scent.[93] They are also attracted by the color and shape of flowers. It can even be argued that color recognition constitutes a chemical interaction, particularly because many of the color pigments in flowers are secondary plant products, such as flavonoids. It is, however, the honeybee's highly developed sense of smell that concerns us here. Honeybees and bumblebees can become conditioned to the scent of flowers that man cannot smell at all. Although bees are guided from a distance by the general form of a plant or flower and especially by its color, the scent of the flower provides the stimulus to alight when the bee is close to it.[92] Although a honeybee usually visits flowers of only one species in a given trip,[94] bees collect pollen from different plants at different times of the day and from many species during a growing season.[92] Obviously many different scents attract the bees.

M. Lepage and R. Boch demonstrated that an attractant in the pollen of certain plants was part of the lipid material and a free fatty acid.[95] Subsequently, C. Y. Hopkins and co-workers identified the fatty acid from bee-collected pollen as a straight-chain trienoic acid with eighteen carbon atoms (octadeca-*trans*-2,*cis*-9,*cis*-12-trienoic acid).[96] This fatty acid was shown to be a strong attractant to honeybees. The pollen extracted for the identification originated from flowers of the legume genera (*Melilotus, Trifolium,* and *Lotus*) and from a species of goldenrod *(Solidago)*.

Readers who desire more information on the crops requiring insect pollination are referred to J. B. Free's *Insect Pollination of Crops.*[92] Free discusses most of the common, and many of the not so common, crop plants that are thought to need insect pollination. He cites evidence of the effects on yields of increasing the numbers of possible pollinators.

Worker honeybees often release a pheromone from the Nassanoff gland when they are near food sources while foraging. This pheromone is extremely attractive to other foragers and appears to stimulate foraging.[27] Interestingly, it is neither colony- nor race-specific. Synthetic Nassanoff pheromone might be useful and economically important in training bees to visit marginally attractive crops.

The use of queen pheromones to stimulate foraging activity in honeybees is another promising application.[27] When queen pheromones were supplied, as synthetic 9-oxodecenoic acid or whole queen extracts, to small test colonies without queens, the colonies showed much greater foraging activity than control colonies. Expendable queenless colonies may possibly be substituted for normal colonies when bees are needed where pesticide exposure is too severe to risk loss or damage to normal colonies.

Because honeybees visit the flowers of many species of plants in response to such a broad spectrum of chemical attractants, there is little specific information on the chemical compounds that initially attract scout bees to particular flowers. For this reason it seems desirable to discuss a few insect pollinators that are restricted to one or a few species of plants, even if the plants are not crop plants. Most of the plants to be discussed below are of economic importance.

In discussing pollination in orchids, L. van der Pijl and C. H. Dodson stated that the sense of smell is virtually the only sense involved in beetle-pollinated flowers, many fly-pollinated flowers, and some bee-pollinated flowers.[97] The enormous variety of odors in orchid flowers provides a practically inexhaustible basis for studying the specificity of insect visitors. According to van der Pijl and Dodson, orchids often deceive visitors by producing putrid odors such as occur in decaying organisms. In other cases the orchids' scents can be detected and distinguished by insects even though the odors are imperceptible to the human nose. Forms of orchid species from different regions sometimes produce different odors, as in the case of the moon orchid *(Phalaenopsis amabilis),* of which only a form from New Guinea produces a strong, sweetish odor. Sometimes special odor glands produce different odors in a single flower. In the orchid *Arachnis flos-aeris* only the tip of the median sepal produces the musklike odor, and it is located far from the column comprising the stamens and the style of the pistil. In many orchid flowers one petal is modified into a landing platform, called the *labellum,* and in *Maxillaria rufescens* the production of the compound vanillin is limited to the parts of the labellum around the food hairs. In some orchids only the petals have an odor, and in others, such as the antelope orchids (*Dendrobium* spp.), only the spirally wound horns produce odors.

In the orchid *Diuris* only specific male-bee pollinators are attracted, and they remain stupefied in the flowers. Members of the subtribes of the orchid family, Catasetinae and Stanhopeinae, produce strong, aromatic odors that have a striking effect on male golden bees *(Euglossa).* Many species in these orchid subtribes attract only specific bee species, and analyses have shown that different chemical compounds are produced by the different orchid species when distinct species of pollinators are attracted.[98] The fragrances of nine species of orchids and a hybrid in the genus *Catasetum* were analyzed by gas chromatography. Five of the nine were found to contain very similar kinds and amounts of chemicals, and those five are pollinated by the same bee, *Eulaema cingu-*

lata (fig. 13). Other bees of the same genus may be attracted, depending on the geographic location. In most instances those orchid species remain distinct only because they have separate geographical distributions. The other four species of *Catasetum* had very different fragrances, from each other and from the five previously mentioned. They are either known or suspected of having different pollinators because of their different flower structures.

Analyses of the fragrances of six species of the orchid genus *Gongora,* including five distinct types of one of the species, demonstrated that all the fragrances have different chemical patterns, including the five types of the one species.[98] All have different species of euglossine bees as pollinators, where they are known, and the others are suspected of having different pollinators also. The fragrances of three species of *Stanhopea* were analyzed, and each fragrance was found to have a different chemical mixture. The *Stanhopea* orchids differed considerably in the color and size of their flowers, and they are pollinated by different bees. They are interfertile in the greenhouse, but the two species that grow together in southwestern Colombia do not hybridize, apparently because their different fragrances attract different pollinators.

These results demonstrate how very important chemical pollinator-insect attractants are in the evolution of at least some of the orchids. More than 10 percent of all species of flowering plants are orchids, and Dodson and his colleagues state that they are so numerous in large part because of the attraction of particular kinds of pollinators.[99] The pollinating agents isolate and prevent hybridization between compatible populations.

Analysis of the floral fragrances of approximately 150 species of orchids from twenty-five genera, mostly pollinated by euglossine bees, indicated that approximately fifty chemical compounds were present.[99] Most of the species produce

Fig. 13. A male bee *(Eulaema cingulata)* scratching inside the lip of a flower of the orchid *Catasetum eburneum.* Reproduced from L. van der Pijl and C. H. Dodson, *Orchid Flowers: Their Pollination and Evolution,* courtesy of the authors.

between seven and ten compounds, but some produce as many as eighteen or as few as three. Sixteen compounds were identified, and ten more tentatively identified. Not all of the compounds are produced with the same frequency. For example, about 60 percent of the species tested produce 1,8-cineole, but less than 5 percent produce methyl cinnamate. The proportions of the different compounds, of course, vary in different fragrances. About 90 percent of the odor of *Stanhopea cirrhata* is due to 1,8-cineole, while only 7 percent of the fragrance of *Catasetum maculatum* is due to that compound.

Field tests were run in various parts of tropical Central America where golden bees are common, using the identified compounds on blotter paper in natural habitats.[99] During a five-day test in central Panama forty-two of the sixty species of golden bees known from that area were attracted to one or more of the known compounds. In later tests twelve of the eighteen species not attracted initially were attracted to the original or to additional compounds. In the initial tests in Panama, 1,8-cineole, methyl salicylate, and benzyl acetate acted as general attractants in pure form, but two of the known compounds, α-pinene and ß-pinene, failed to attract any bees. Of the compounds identified, 1,8-cineole has been the most effective general attractant, except in western Mexico, where eugenol was as effective generally and was more effective with particular kinds of bees. Myrcene was the only compound to attract female golden bees, and only two individuals were attracted in the tests in the three Central American countries. Many species of bees were attracted to several compounds, but some were attracted to only one.

A combination of two attractants changes the attraction potential of the total fragrance and often fails to attract species that are drawn to one of the compounds in pure form. The addition of still a third compound sometimes reduces further the number of species attracted. The selectivity of the bees responding to the mixtures is very important, because some combinations of compounds will attract only one or a few species of golden bees. The differential production of fragrance components by different species of orchids can limit the number of bee species attracted.[99] Preliminary research with certain species of orchids that are adapted to moth pollination indicates that species-specific fragrances are produced in these orchids also.

Ironically, the euglossine bees, which are so important in the pollination of many orchids, obtain absolutely no food from them. The orchid flowers have no nectaries, and pollen is not taken for food. Moreover, only male bees visit the flowers. Careful observations and subsequent analyses have demonstrated that these bees carefully choose a flower, be-

106

cause of its appropriate fragrance, and rub the lip of the flower with special brushes on their front feet. They then take off again and hover in front of the flower while they deposit the odor substance in cavities in their swollen hind legs.[99] This process is repeated on the same flower or other flowers with a similar fragrance. It was first thought that these males bees collected the fragrance components in particular combinations to attract females of their own kind. Experiments demonstrated, however, that the fragrances attracted only male bees.[100] Subsequently it was found that males that have collected sufficient amounts of the fragrance material do curious flights and buzzing rituals that attract other male bees of the same species. Eventually female bees are attracted to the swarm, probably by the accumulation of brilliantly colored bees, and mating takes place. Thus the chemical attractants serve a function in the perpetuation of the golden bees as well as the orchids that they pollinate.

Representatives of many families of small moths visit flowers. These nocturnal moths approach the flowers chiefly by flying upwind, which suggests they are flying along an odor trail. If netting is placed over the flowers, the moths land on the netting and extend the proboscis—an indication that they react to the flowers' odor, since feeling and sight are eliminated. According to the researcher N. B. M. Brantjes, mosquitoes show a similar behavior. Both insects visit the same flower species for drinking and often pollinate them.[101] Mosquitoes were shown to be attracted by odor also, because they were lured by flowers hidden in boxes. Other small Diptera visit and pollinate the same flowers visited by some of the small moths.

The great families of large moths—the hawk moths (Sphingidae) and owlet moths (Noctuidae)—are also attracted to flowers by their odors.[101] In an experimental cage where moths were in flight, both hawk moths and owlet moths demonstrated an abrupt change in behavior after in-

107

troduction of a flower odor. In an odorless atmosphere the moths flew slowly around, but after the introduction of the attractive odor, their flight pattern changed to include many sudden drops and slow rises. Such a pattern has been termed "the seeking flight." In the absence of flower odor the moths avoid flying into colored objects, indicating that they can see the objects but are not attracted to them. Since the moths start to approach the colored objects after the odor is introduced, the flower odor is clearly an arrestant. The hawk moths appear to react to specific odors, whereas the owlet moths react to a broad spectrum of flower odors. The numbers of landings on test-paper strips containing flower odors have been shown to be proportional to the concentrations of the odors—an indication that the moths use concentration differences for orientation.

Odor differences, either qualitative or quantitative, on the surface of some flowers serve as nectar guides to moths.[101] A structure, called a ligula, on the flower of snowy campion *(Silene alba)* directs the proboscis of the moth toward the flower opening. If the ligula is removed, the moth will scan the entire surface of the petal with its antennae and proboscis tip to find the opening in the floral tube.

Many species of the arum plant family (Araceae) produce rather putrid odors that attract various pollinator insects that normally feed on decaying organic matter. Genuine species of this family have no showy sepals and petals. The flowers are crowded on a fleshy stalk, the spadix, which is usually subtended by a large smooth bract, the spathe. The tip of the spadix, called the appendix, has no flowers attached and is very smooth. It produces so much heat that temperatures on its surface can be as much as 22°C higher than the surrounding air. The heat aids in the evaporation of odor-producing chemicals, such as amines, ammonia, indole, or skatole. Those types of compounds and mixtures of them can produce a large variety of odors which are highly specific in attracting pollinators. The arum *Helicodiceros musci-vorus* is pollinated almost exclusively by large flies; *Arisarum proboscideum,* which smells like a mushroom at the appropriate stage, is pollinated by fungus gnats; and *Dracunculus vulgaris* is pollinated by beetles.[102]

There are many other examples of plant-insect chemical interactions in insect pollination, but those discussed above should suffice to demonstrate the very great importance of such interactions in agriculture and in biology in general.

CHAPTER 6

Possible Uses in Insect Control of Chemical Interactions Between Organisms

THE pace of research on plant-insect chemical interactions, pheromones, and insect hormones accelerated greatly in the 1970s. Scientists had two main goals in mind:

(1) To increase our basic understanding of such biological phenomena.

(2) To decrease our dependency on pesticides.

Greater knowledge of the biology of the insects that we need to control will also increase the efficiency and effectiveness of our use of pesticides when they must be employed in control of pests.

Two basic problems have been encountered in the use of the insecticides that were developed in the past:

(1) Insecticides are toxic not only to the pests but also to many other insects and animals as well.

(2) Target insects rapidly develop resistance to them.

Some have been persistent in the environment also and become concentrated through food chains. Obviously we want specific control methods for target insects that will continue to be effective because the insects cannot acquire resistance to them. It has been estimated that only about 0.1 percent of all insect species are harmful as pests in agriculture or as vectors in the transmission of human and animal diseases.[1]

NATURAL INSECTICIDES IN PLANTS

The tremendous destructive potential of plant-eating insects and pathogens has failed to prevent green plants from dominating most of the terrestrial surface of the earth. Increasing evidence suggests that the survival of plants is due largely to their own defensive strategies, which have coevolved through time in relation to the attack strategies of herbivores and pathogens. Plant defenses can be structural or chemical, but chemicals seem to be especially important.[2]

Plant-produced chemicals repel insects, deter feeding and oviposition, and have delayed effects on growth and reproduction. From the discussion in Chapter 5 it is evident that these chemicals could be effective against insect pests. Certainly some of the plant-produced chemicals that are toxic to some or all insects will play a defensive role in insect control, and some of these also serve as feeding deterrents. The many plant-produced chemicals that act as hormones in insects in connection with molting and maturation[3] will be discussed briefly in this chapter. These could be considered along with the plant-produced chemicals that have marked delayed effects on growth and reproduction, but we know that they are, in fact, similar or identical to the actual hormones produced by certain insects.

It is noteworthy that plants in natural ecosystems are dependent entirely on their own defenses against insects and other herbivores—an indication of how effective the natural defenses can be. Many of the chemical compounds involved, such as tannins and alkaloids, are bitter-tasting, and many are toxic to mammals and other animals. Crop plants have often been selected and bred for lower concentrations of such compounds. The not-too-surprising result, in light of our present knowledge of natural chemical defenses, is that many crop plants are relatively susceptible to grazing by insects. Since many crop varieties are relatively uniform genetically, virtually all individuals of a given variety may be equally susceptible to insect predations. Obviously crop plants are generally selected and bred for particular structural characteristics, and these changes can lower the plants' defenses against insects. Also, large groups of similar plants are easier for insects to find than the relatively isolated individuals that often occur in natural ecosystems. All of these factors have led, at least in part, to our increased dependence on commercial insecticides in agriculture.

I pointed out in Chapter 5 that most crop plants have at least some resistance to some insect pests. Often varieties or related species can be found with considerable resistance to selected pests. These can sometimes be used in breeding programs to make varieties with other desirable characteristics resistant, or at least more resistant, to the same pests. The Kansas Agricultural Experiment Station has led this kind of research in the United States.[4] The station has approved for distribution to farmers eleven crop varieties that are resistant to various insects, including varieties of sorghum, alfalfa, corn, barley and wheat.

Development of wheat varieties resistant to the Hessian fly *(Phytophaga destructor)* has been very important economically. About ten million acres are now planted with such varieties each year in the United States.[4] As of 1968, $15 million in increased yield was being made in Kansas alone because of these resistant varieties. The value of increased yields in all of the United States in 1964 was calculated at $238 million. Wheat varieties resistant to the wheat-stem sawfly *(Dolerus* spp.) have produced similar profits.[4] Although insecticide control of this insect has not been economically feasible, the use of two resistant varieties, Rescue and Chinook, allows the profitable production of wheat on almost two million acres in Canada and over one-half million acres in the United States. The wheat-stem sawfly previously caused crop losses up to 80 percent on the Great Plains of Canada and the United States.

Use of resistant varieties often increases the quality of a crop as well as the yield. Moreover, the resistant plants often decrease the insect population in susceptible neighboring crops as well. Since there is no direct cost to the farmer, the use of resistant varieties reduces costs of production where the margin of profit is small. All other control measures require the continuous involvement of man before, during, and even after crop production, whereas resistance enables plants to defend themselves against pests, or against a level of predation harmful to the crop. The relative success of other control measures is often determined by the precision with which they are used, but resistance is not so subject to external variations. Another distinct advantage of resistant plants is that insects feeding on them are often less vigorous and more easily killed by adverse environmental conditions.[4] They can be controlled, if necessary, with smaller amounts of insecticides. Predation on the pests by other insects has been shown to be greater on some resistant plant varieties, resulting in better biological control. Thus resistant plants are more compatible with supplementary control measures.

The more effective control measures against plant diseases caused by bacteria, fungi, and viruses have involved selection and breeding of crop varieties resistant to the pathogens. Dr. R. R. Nelson estimated that by 1973 more than 75 percent of the current agricultural acreage in the United States had been planted with varieties resistant to one or more of the plant diseases.[5] It is surprising therefore that use of resistant plant varieties in the management of insect pests has been relatively limited.

In Chapter 2, I discussed several plants that have been known for hundreds of years to have insecticidal properties. In most instances, however, the active compound, or compounds, was not known until relatively recently. It is appropriate at this point to discuss some of the kinds of chemical compounds in plants that have insecticidal activity.

Most works on so-called poisonous plants are concerned with their effects on human beings and domestic animals. It is important to keep in mind that a compound that is toxic to one type of insect is not necessarily toxic to other insects or other animals. Many of the supposedly toxic plants are eaten avidly by at least a few insects without harm to the insects, though, of course, some are toxic to many insects. Conversely, rotenone, which is very toxic to insects, has relatively low toxicity to mammals. Some insects are able to detoxify certain compounds that kill other insects. Many of the secondary plant products can be insecticidal to at least some insects in sufficient concentrations.

Thus any system designed to classify toxins in plants has built-in defects. At the very least, the type of animal needs to be specified. John Kingsbury in his *Poisonous Plants of the United States and Canada* gives a rather thorough discussion of poisonous principles in plants, chiefly in relation to livestock and humans.[6] The chemistry of the compounds is discussed for the reader who wishes more details. E. A. Bell has reviewed research on toxins in seeds and gives a classification of toxins that correlates well with the major categories of Kingsbury.[7] The eight major categories of toxins in seeds listed by Bell were (1) phytohemagglutinins, (2) enzyme inhibitors, (3) polysaccharides, (4) cyanogens, (5) saponins, (6) alkaloids, (7) unusual amino acids, and (8) miscellaneous toxins.

Phytohemagglutinins. Most of the phytohemagglutinins are carbohydrate-containing proteins, but some are lipid-containing proteins, and at least one is a protein without lipid or carbohydrate.[7] All cause clumping of red blood cells, which is the reason for their name. The first to be discovered was ricin in castor beans (spurge family). Now they are known to be widely distributed in the plant kingdom, even in some microorganisms, but the best known are in the spurge and legume families. Some other well-known phytohemagglutinins are abrin in rosary pea *(Abrus precatorius),* crotin in croton *(Croton tiglium),* robin in the bark of black locust *(Robinia pseudoacacia),* and soyin from soybean. G. C. Toms and A. Western give a long list of the phytohemagglutinin compounds that have been found in the legume family.[8]

Many of the phytohemagglutinins are toxic to higher animals, but their toxicity is not necessarily related to the clumping effect on red blood cells.[7] At least some of these compounds are toxic to some insects, as demonstrated by D. H. Janzen and his colleagues,[9,10] who found that a phytohemagglutinin from black bean was lethal to larvae of the southern cowpea weevil *(Callosobruchus maculatus)* in certain concentrations.

Enzyme Inhibitors. S. W. Applebaum and Y. Birk found an inhibitor of flour-beetle larval protease in peanuts, soybeans, chick-peas, and other legume seeds.[11] They suggest that selection or breeding of legumes containing higher concentrations of this enzyme inhibitor may help prevent damage to the seeds in storage. Inhibitors of amylases (starch-digesting enzymes) have been found in the seeds of at least twenty species of legumes and in wheat and rye.[7] The inhibitor found in wheat is a specific inhibitor of the alpha-amylase produced by larvae of the seed-eating beetle *Tenebrio molitor.*

Polysaccharides. A complex polysaccharide (a molecule made of many sugars) appears to be the basis of the resistance of haricot beans *(Phaseolus vulgaris)* to the larvae of the seed beetle *Callosobruchus chinensis.* Normally, this compound constitutes about 1 percent of the dry weight of the seeds. This concentration is not inhibitory to another seed-eating beetle, *Acanthoscelides obtectus,* which is a major pest of stored haricot beans, but higher concentrations impart resistance to this beetle also. S. W. Applebaum and Y. Birk suggest that it would be logical to breed for higher concentrations of this polysaccharide in haricot beans.[11]

Cyanogens. Over one thousand species of plants are known to produce hydrogen cyanide (HCN), generally as a result of tissue damage, which causes a cyanogenic compound present in the plant to come into contact with an enzyme capable of releasing HCN from it.[7] The known cyanogens are either cyanogenic glycosides or cyanogenic lipids. Amygdalin is such a glycoside. It was very much in the news in 1980 and 1981 because of its controversial use in the treatment of cancers. It occurs in bitter almond and in the kernels of other members of the rose family, such as cherry and peach.[7] It also occurs in the roots, leaves, and stems of several of those species.[12] Digestion of amygdalin produces a sugar (gentiobiose), benzaldehyde, and HCN. There are many other cyanogenic glycosides in many kinds of plants. The sugars and the aglycones (nonsugar portions) are quite variable, but all cyanogenic glycosides produce hydrogen cyanide, which is toxic to many animals, including many insects.

Cyanolipids have been found in seeds of several species of the soapberry family (Sapindaceae) and in one species of the borage family (Boraginaceae).[13] All do not produce HCN on hydrolysis. Nevertheless, seed oils from the goldenrain tree *(Koelreuteria paniculata)* and from soapberry (*Sapindus*

spp.), both of which contain cyanolipids that do not produce HCN, are toxic to cowpea-weevil larvae in concentrations of 0.1 to 1.0 percent in food.[10]

Saponins. Saponins are glycosides in which the nonsugar
portion of the molecule is a sterol or a triterpene.[7] These
compounds, which occur in seeds and other plant parts of
many species, are highly toxic to cold-blooded animals. Commonly the saponin content varies with the part of the plant,
the stage of growth, and the season. These compounds occur
in relatively high amounts in the vegetative parts of corn
cockle *(Agrostemma githago),* tung tree, beech, English ivy,
yellow pine flax *(Linum neomexicanum),* alfalfa, pokeweed
(Phytolacca americana), bouncing bet (*Saponaria* spp.), and
coffeeweed (*Sesbania* spp.).[6] The seeds of most of the cultivated legumes contain saponins, several of which are toxic
to the larvae of the seed beetle *Callosobruchus* and the flour
beetle *Tribolium.*[11]

Alkaloids. The *alkaloids* are basic compounds, as the name
indicates, and they contain one or more rings with a nitrogen atom.[7] Some well-known ones are nicotine, atropine, qui-
nine, morphine, strychnine, reserpine, and colchicine. Alkaloids are widely distributed in the plant kingdom as is demonstrated by the survey of S. J. Smolenski and his colleagues,[14]
who tested 5,550 species and found alkaloids in 1,526. Many
of the alkaloids are very toxic to vertebrates and to insects,
as the list above indicates. Nicotine has been used widely
as an insecticide for several hundred years, as described in
Chapter 2. D. H. Janzen and his colleagues tested eleven
alkaloids against the larvae of cowpea weevils and found all
to be lethal at concentrations of 0.1 to 1.0 percent in the
diet.[10] All but two were toxic at the low level.

Unusual Amino Acids. Over two hundred amino acids have

been found in plants which are never found as normal constituents of proteins nor as metabolic intermediates.[15] Many of these can act as antimetabolites and thus as toxins to some organisms. In fact, some were shown to be toxic to most of the animals against which they were tested and to microorganisms and some higher plants.[7] Janzen and his colleagues tested a large group of nonprotein amino acids against larvae of the cowpea weevil and found that most of those tested were lethal.

Miscellaneous Toxins. Several additional kinds of compounds, mentioned often in Chapters 3 and 5, are insecticidal to at least some insects, in appropriate concentrations. These include mustard oil glycosides, tannins, phenolics, flavonoids, free amines, and others.[7,10,16,17]

Naturally Occurring Insecticides Used Commercially

Nicotine. Nicotine, as we have seen, is one of the oldest and most widely used commercial insecticides. Nicotine has been isolated from at least eighteen species of the genus *Nicotiana* (from which it derives its name). It is found in a few species in other plant genera as well.[18] In the United States it is still produced commercially from cultivated tobacco *(N. tabacum)*. It is extracted from the woody stems and leaf midribs, which are not suitable for smoking or chewing. Another tobacco species *(N. rustica),* of higher alkaloid content, is cultivated in the Soviet Union, Germany, Kenya, and Hungary as a source of nicotine. According to I. Schmeltz, *N. rustica* has up to 18 percent nicotine versus about 6 percent in *N. tabacum*. Since tobacco breeders have been selecting for lower nicotine concentrations in *N. tabacum, N. rustica* is now the chief commercial source.[19]

Most of the nicotine used in the United States is sold in the form of a sulfate in a water solution containing 40 percent of the alkaloid. Free nicotine is available also, but is not so widely used in spraying because it evaporates too quickly and is much more toxic to human beings than the sulfate.[18] The spray is diluted greatly for use, as concentrations of 0.05 to 0.1 percent nicotine are sufficient to kill most aphids and other soft-bodied insects.

In 1929, Russian chemists isolated an alkaloid related to nicotine from an Asiatic shrub, *Anabasis aphylla,* belonging to the goosefoot family (Chenopodiaceae). They named it anabasine, and it is an isomer of nicotine. It has the same proportions of carbon, hydrogen, and nitrogen and is like nicotine in its chemical, physical, and insecticidal properties.[18] The Soviets have produced anabasine sulfate commercially from this shrub.

Rethrins. Pyrethrum is a powder made from the dried flowers of *Crysanthemum cinerariaefolium.*[20] It has been used as an insecticide since ancient times and was introduced into Europe in 1828, into the United States in 1876, and then into Japan, Africa, and South America. Its use spread rapidly because, in the doses necessary to kill household insects, it is nontoxic to humans and domestic animals.[18] In about 1916 kerosene extracts of pyrethrum flowers appeared on the market and were used widely as sprays against flies and mosquitoes. Later the active compounds were extracted and identified. The chief insecticidal components were named pyrethrin I, pyrethrin II, cinerin I, cinerin II, jasmolin I, and jasmolin II.[20] Collectively they are called rethrins. Highly purified, standardized concentrates are now available.

The rethrins rapidly paralyze insects such as houseflies. Therefore they are often used with slower-acting insecticides that have a longer residual effect. The rethrins deteriorate relatively rapidly on exposure to light and air, so that various materials are added to prevent their decomposition. Any reader who desires more information on pyrethrum is referred to the book edited by John Casida, *Pyrethrum — The Natural Insecticide.*[21]

Rotenoids. Natives in many tropical countries use water extracts of various species of legumes to catch fish, and some of the extracts are also used to kill insects on human beings and domestic animals.[18] Tuba root from *Derris* and cubé root from *Lonchocarpus* became common commercial insecticides about 1930. Their active ingredients are rotenone and related compounds called rotenoids.[22] The original method of using tuba root, developed by Chinese gardeners in the Malay Peninsula, was to crush the fresh derris roots in water and spray the liquid on the plants. Now derris and cubé are used mostly as dusts. The dried roots are ground and mixed with an inert substance to give a final rotenone content of about one percent.[18]

Ryanodine. An alkaloid, named ryanodine, from *Ryania speciosa,* a tropical shrub native to Trinidad, is highly specific against the European corn borer.[18] The highest concentration occurs in the roots, but the source of the commercial insecticide is the stem wood. Thus the entire plant is not destroyed, because it regenerates a new top. The finely ground plant material can be used as a dust or suspended in water as a spray.

Unsaturated Isobutylamides. Several plants produce the insecticides that are classified by chemists as unsaturated isobutylamides. The roots of *Anacyclus pyrethrum,* a North African plant in the composite family, contain one of these compounds, named pellitorine. The roots of another composite, *Helliopsis longipes,* have been used as a native insecticide in Mexico and have been found to contain a related amide compound, named affinin. Another species of *Heliop-*

sis (H. scabra) has been found to contain two amides, termed scabrin and heliopsin, and both are powerfully insecticidal. The roots of at least two species of the purple coneflowers *Echinacea angustifolia* and *E. pallida* contain another related insecticidal amide, named echinacein. Prickly ash *(Xanthoxylum clava-herculis)*, a shrub in the rue family, contains appreciable amounts of two similar amides in its bark, which have been named herculin and neoherculin.[23]

Minor Insecticides of Plant Origin. D. G. Crosby has discussed what he terms minor insecticides of plant origin.[24] Among those he lists are compounds from *Quassia amara* in the Simarubaceae and from false hellebore *(Veratrum album)* in the lily family. Materials from both of these plants have been used as insecticides for a very long period of time, as indicated in Chapter 2. The wood and bark of quassia contain an insecticide named quassin, and the bark and wood of the tree of heaven *(Ailanthus altissima),* which is in the same family, contain a related insecticide, named neoquassin. The dried rhizomes of false hellebore contain several insecticidal alkaloids, and some of its essential oils have insecticidal activity. These, however, are used primarily in conjunction with other insecticides in commercial preparations.[18,25]

USES OF CHEMICALS THAT AFFECT INSECT BEHAVIOR

We have seen that many phases of insect behavior are stimulated and controlled by chemicals: the location of food, oviposition sites, and sexual partners; feeding; copulation; egglaying; and others. In addition to selecting and breeding plants for the production of chemicals that will protect the plants from insect predation, it seems logical simply to use the chemical compounds or synthetic compounds with similar effects to control insect behavior in ways that will protect crop plants.

Natural sex attractants and mating stimulants are produced by many pest insects. These are often effective in very small concentrations. For example, one virgin female pine sawfly in a cage attracted over 11,000 males.[4] Many such compounds have been identified and synthesized, and many other chemical lures have been synthesized.

Chemical lures, or attractants, may be used to attract and trap insect pests; to lure them to contact with poisons, chemosterilants, or pathogens; or to mask the location of mates by saturating the environment with synthetic sex pheromones.[4] Such techniques promise highly specific control of pests with few or no ecological side effects. Already the use of chemical lures to assess population densities of pests can be extremely helpful in integrated control programs. Later in this chapter the use of chemical attractants for insect pest control in crops will be discussed further.

Obviously, chemical repellents and chemical feeding or oviposition deterrents also offer intriguing possibilities for behavioral control of insect pests.

Repellents

Certain plants have been used since antiquity to keep insect pests away from clothes closets and other storage areas and to keep certain pests off the human body (see Chapter 2). In such cases use is made of the natural repellents in the plant material. The natural insecticides in certain plants have been similarly exploited, and until the past several decades they were the chief source of chemicals for insect control.

The use of certain plant materials as insect repellents led directly to the extraction and identification of several essential oils that are fairly efficient repellents. Two examples are oil of citronella and oil of camphor. Oil of citronella was apparently the most widely used mosquito repellent during about the first one-third of this century. This substance is extracted from a grass *(Andropogon nardus)* and contains geraniol as its primary component with lesser amounts of citronellol, citronellal, borneol, and other terpenes.[26] Citronellol and citronellal are considered to be the principal mosquito repellents in the oil.

Between 1935 and 1955 research for insect repellents in the United States was directed toward the production of synthetic chemicals for the protection of humans.[26] United States Department of Agriculture scientists have collected, synthesized, and tested more than 20,000 chemicals as repellents against a variety of arthropods.[27] Though many of the compounds were effective, only a few were safe to use on skin. According to Ruth Painter, there were four standard repellents for biting arthropods in general use before World War II: oil of citronella, dimethyl phthalate, indalone, and Rutgers 612 (now called 612). The most successful, Rutgers 612 (2-ethyl-1,3-hexanediol) is still in use. In subsequent screening USDA scientists discovered several compounds that are of practical value as repellents. One of these, deet (N,N-diethyl-m-toluamide), is the best single repellent available today.[27] It is effective against many insects, including mosquitos, chiggers, ticks, deerflies, sand flies, biting gnats, and land leeches. In recent years the emphasis has been on new formulations to make the repellents more acceptable and easier to apply. Deet is now available in the original liquid form (50 percent solution), in a cream base, a foam-type lotion, and a spray.[26] It is effective when applied to the skin or to clothing. Readers who desire more details on the chemistry of repellents, and on repellents other than those mentioned here, should refer to the articles by Painter[26] and by Metcalf and Metcalf[28] cited in the Notes.

Beekeepers have always needed a repellent at harvest time. In the old days smoke was widely used and it is still used today. Phenol has been extensively used also, but propionic acid, benzaldehyde (artificial oil of almonds), propionic anhydride, and acetic acid have been recommended as replacements for phenol because they are safer to handle. A. W. Woodrow and his colleagues reported that propionic anhydride and propionic and acetic acids are all ideally suited to remove bees from the honeycomb because they are good repellents, are nonpersistent, and do not contaminate the honey.[29] Loss of bees from insecticide poisoning has been a major problem in the bee industry since the widespread use of synthetic organic insecticides began,[26] and there have been many attempts to keep bees away from dangerous areas by incorporating a bee repellent in the insect control program. Although none of those attempts has been successful, the possibility for success still exists.

A good insect repellent for use on domestic livestock would boost milk and meat production, but a satisfactory chemical for this use has not been found.[27] According to R. L. and R. A. Metcalf, sprays of synergized pyrethrins can protect livestock against all biting flies if applied twice daily. This is time-consuming and expensive, of course, and would probably not be practical for most farmers.

According to M. Beroza, no one has developed a satisfactory insect repellent for use on agricultural products.[27] A volatile repellent does not last long, and frequent spraying of a repellent over a long growing season is not economically feasible. Nonvolatile materials are actually feeding deterrents if they are effective at all, and not repellents, and many of the so-called repellents described in the literature are feeding deterrents. These repellents, or deterrents, will be discussed below. The Metcalfs point out that the use of foliage repellents offers few advantages in pest-management programs and some distinct disadvantages.[28] They claim that the use of repellents simply directs insect attacks to untreated crops, and that the chemicals that are effective provide about the same degree of environmental hazard as conventional insecticides without any decreases in pest populations. It is still possible, of course, that practical new chemical repellents may be discovered.

Creosote, or 4,6-dinitro-o-cresol, is a very effective repellent against the migration of chinch bugs *(Blissus leucopterus leucopterus)* from small grains to corn. If applied on the soil between the crops,[28] it provides good crop protection with minimal environmental contamination or disturbance.

Deterrents

According to the Metcalfs, the most successful feeding deterrent on foliage is still Bordeaux mixture, a fungicide developed in France almost 100 years ago.[28] It is produced from copper sulfate, hydrated lime, and water in a 6-10-100 mixture and acts as a feeding deterrent to flea beetles, leafhoppers, and the potato psyllid *(Paratrioza cockerelli)*.

Several compounds have been developed for mothproofing clothes and other materials, and they are effective because they prevent the feeding of the moth larvae. According to Donald Wright, such a mechanism was not recognized as a potential method of insect control in agriculture until Compound 24,055 (4'-[dimethyltriazeno] acetanilide) was introduced in 1959. The class of compounds to which Compound 24,055 belongs, the triazenes, is the only class of feeding deterrents to be tested extensively outside the laboratory. In extensive field tests in the United States and Canada it was demonstrated that at practical rates of application Compound 24,055 is not toxic to most insects but inhibits the feeding of most of the surface-feeding, chewing species. Chewing insects that feed under the treated surface, such as borers, earworms, and codling-moth larvae, were not affected, because they did not have to encounter the material after the first bite or so. This compound is relatively nonspecific, and many diverse pests are affected. Nevertheless, the pests that it controls did not represent a large enough market in 1961 to justify the costs involved in registration for use at the required rates of application. In 1964 it was tested again for possible control of the cotton bollworm, after this pest had developed resistance to insecticides. Again the material was found to be only marginally effective at competitive rates of application.

The organotins are the only other class of feeding deterrents that has been studied relatively broadly. Several triphenyltins have been shown to have pronounced feeding-deterrent activity. Brestan (triphenyltin acetate) has been reported to be effective in field tests against the cotton leafworm, the Colorado potato beetle, the potato-tuber moth, and the caterpillar of *Agrotis ypsilon.*[30]

The carbamates, such as Sevin, are used primarily as insecticides, but several investigators have reported that they are also feeding deterrents against some insects. Foliar applications of several thiocarbamates were reported to inhibit feeding of Mexican bean beetles, Colorado potato beetles, and Japanese beetles. Many of the phenylcarbamates with alkyl or alkoxy substituents on the molecule prevent the feeding of the salt-marsh caterpillar at doses about one-tenth of the lethal rate, and similar results were reported for the boll weevil. Baygon (o-[isopropoxyphenyl]-N-carbamate) has been shown to be a systemic antifeedant against the boll-weevil at rates from 40 to 100 parts per million.[30]

Pyrethrum is a feeding deterrent to the biting flies *Glossina* and *Culicoides,* in addition to being toxic to many insect pests. Several plant growth regulators are relatively effective in inhibiting the feeding of the cotton leafworm. These include phosfon, cyclocel, B-nine, and carvadan. Phosfon is most effective. A leaf dip in a 4,000 ppm solution inhibited 89 percent of feeding on cotton. Forty parts per million in water protected cut chrysanthemums from damage, and in yield tests with peppers and peanuts, cotton leafworms were controlled with foliar sprays of 1200 and 400 ppm, respectively.[30]

Unfortunately, feeding deterrents are impractical at the present time for agricultural crops. The cost of application is too high, and so far the feeding deterrents tested have been effective only on the tissues where they are deposited. Growing plants constantly produce new tissue, and any tissue that grows after spraying is not protected. Development of more effective and cheaper feeding deterrents may make the use of them feasible at least in truck farming and gardening. One distinct advantage of such compounds is that they are not harmful to pollinators.[31]

John Capinera and Frank Stermitz point out that, to be practical, a feeding deterrent should have the following principal properties: it should be relatively persistent, it should be translocated through the plant to untreated parts, and it should have no harmful effects on nontarget organisms. They state further that some natural plant products, such as azadirachtin, a triterpenoid from the neem tree *(Azadirachta indica),* are persistent, move through the treated plant, and have specificity. They suggest that alkaloids may be useful in this regard. Zanthophylline is an alkaloid recently isolated from *Zanthoxylum monophyllum.* They tested it as a feeding deterrent against the range caterpillar *(Hemileuca oliviae)* on corn, the migratory grasshopper *(Locusta migratoria)* on barley, the alfalfa weevil *(Hypera postica)* on alfalfa, and the greenbug on barley.[32] In choice tests the first three insects reduced their feeding on the zanthophylline-treated plants, but only the range caterpillar was inhibited in no-choice tests. Greenbug host preference was not affected by zanthophylline.

Gy. Sáringer tested fifteen rather diverse compounds against egg laying by the poppy beetle in poppy fruits.[33] Thirteen of the compounds inhibited egg laying by this beetle at least to some extent. Most had pronounced deterrent effects.

In Chapter 5 it was pointed out that aqueous extracts of chestnut leaves inhibit egg-laying by the sugar-beet moth on its host plant. Thus it appears that natural plant products may be found that will effectively prevent or reduce the reproduction of at least some insect pests on crop plants.[34]

Attractants

Chemical attractants are very important determinants of insect behavior because they direct insects in their searches for food, the opposite sex, and a place to lay their eggs. The sex attractants are emitted by one member of a species to attract a mate for mating and propagation. These pheromones usually attract only their own species—and even in tiny amounts they often attract mates from relatively great distances.[35]

There was rapid progress in the identification of sex attractants during the 1970s. Whereas only a few had been identified before, by 1980 the attractants of several hundred different insect pests had been identified.[35] Sex pheromones have three chief uses in pest control:

(1) Detection and survey of insect species
(2) Mass trapping
(3) Disruption of the odor-guidance system that brings the sexes together for mating.

Detection and survey traps are widely used in pest-management programs. The capture of particular pests that have been lured into traps indicates that they are present, and the numbers captured can determine whether control methods are necessary. Such surveys can cut down markedly on the amounts of insecticides used in control. Also, on occasion, prompt action can prevent the spread of pests into uninfested areas.

As of 1976 about 17,000 traps baited with synthetic lures were being maintained by the USDA across the southern periphery of the United States to detect any accidental importation of three highly destructive subtropical pests.[35] The three pests of concern are the Mediterranean fruit fly *(Ceratitis capitata)*, the melon fly *(Dacus cucurbitae)*, and the oriental fruit fly *(Dacus dorsalis)*. This early-warning system has saved many millions of dollars by detecting minor infestations and eliminating them before they spread. Traps are also distributed around several ports of entry for the same purpose. In 1974 more than 100,000 lure-baited traps were distributed in the area east of the Mississippi River to detect infestations of the gypsy moth.[35]

In 1956 the Mediterranean fruit fly was found to infest about 1 million acres within the state of Florida. With the help of attractant-baited traps to locate the insects, the government was able to eradicate the pests in 1957, using insecticides.[27] The oriental fruit fly was eradicated from Rota Island in the Mariana Islands of the Pacific Ocean by luring the male flies to a poison-coated surface with an artificial attractant, methyl eugenol.[4]

It takes a lot of research and field trials to learn how to use the sex attractants properly for the control of different pests. The responses of the insects to these compounds vary from species to species just as do the attractants produced by different species. Morton Beroza points out that many parameters need to be considered for successful use of lure-baited traps.[35] Some of these are trap design, trap height, trap placement, trap durability, the type of lure dispenser, the position of the lure dispenser in the trap, the emission rate of the lure, lure stability and quantities, the duration of effectiveness, the ratio of components in the lure (if more than one compound is involved), the effect of the host crop, the effective distance of attraction, the time of insect response, the cost of the trap, and the means of eliminating the insects that are trapped (adhesive, insecticide, or whatever). Gypsy-moth traps that were tested at heights up to four meters caught best between ground level and two meters. The greatest catch of oriental fruit moths *(Grapholitha molesta)* was at one meter above the ground. The catch was somewhat less at two meters; only a few moths were caught at three meters; and only a very few were caught at ground level. Catches of the pecan-bud moth *(Gretchena bolliana)* were very low near the ground and increased with trap heights up to 9.1 meters, the greatest height tested.

Certain insects can be controlled by mass trapping, but usually only when the infestations are relatively light. If the population is too high for effective use of traps, it can be lowered by use of insecticides. After that traps may prevent another increase in the population. At lower population levels trapping may eradicate a target species.

Dispersal of a sex attractant widely over an area may disrupt the guidance system of the target species and prevent males and females from getting together. This method has special appeal in that reproduction of the insect pest may be suppressed with a harmless chemical that has no appreciable effect on the environment or on any other organisms besides the target pest. Tests on the gypsy-moth sex attractant and on eight other sex attractants have shown that they have a low order of toxicity. One nanogram of disparlure, the sex attractant of the gypsy moth, suitably formulated with a keeper, remained effective for three months in the field. Moreover, this sex attractant is available in relatively large amounts; 1,200 pounds were purchased in 1975.[35] Like trapping, the air-permeation or disruption method becomes more efficient as the population of the target pest declines. If the population is too high, males may find females by

130

chance, and this possibility increases as the numbers of the pest insect increase.

Some examples of successful control of selected pests by means of sex attractants will demonstrate the potential of these pheromones. The production of cotton dates back several thousand years, and many insects have become adapted to living and feeding on this plant. Throughout the world 1,326 species of insects have been found on cotton. About 100 species of insects and mites attack cotton in the United States alone, though fortunately only a few of these ever become serious pests. Most of the serious damage is done by insects that attack the flowers and fruits (the bolls). The USDA Crop Reporting Board has reported that one bale of cotton lint is lost to insect pests for every eight bales produced. Moneywise, the average annual loss between 1951 and 1960 was $500 million in the United States.[36]

One very serious pest to cotton is, of course, the boll weevil, which occurs from Texas east to the Atlantic Ocean. In 1976, P. A. Hedin and his colleagues reported that $70 million was spent annually for boll-weevil control in the United States, and the weevil still causes crop losses of $200 million each year. More than three-fourths of all the losses of cotton to insects in the United States have been attributed to the boll weevil,[37] and about one-third of the insecticides used in agriculture in the United States each year are used for boll-weevil control. Thus alternate or substitute control measures are badly needed.

In 1962, W. H. Cross and H. C. Mitchell demonstrated conclusively that the male boll weevil produces a wind-borne sex attractant to females.[38] Later it was shown that this pheromone acts as an aggregating pheromone, or arrestant, for both sexes. This pheromone was shown to consist of four terpenoids in a ratio of 6:6:2:1/I:II:III:IV, where the Roman numerals designate the terpenoids.[37] This sex attractant, named grandlure, has been synthesized in the laboratory and tested widely. It acts as an arrestant for both sexes in addition to being a sex attractant for the females.[36]

D. D. Hardee and his colleagues distributed wing traps baited with males in thirty-four cotton fields in west Texas, which totaled 1,542 acres.[39] The caged males were replaced in the traps once each week from April 5 to July 1 and twice each week through August 22. When the traps were removed from the fields on August 29, no boll weevils and no feeding- and oviposition-punctured squares had been found in seventeen of the thirty-four fields, and live boll weevils had been found in only seven of the seventeen infested fields. Moreover, populations of boll weevils did not develop to injurious levels in the infested fields. Traps placed around all thirty-four fields, however, captured boll weevils in low to moderately high numbers.

Initial field trials with several formulations of grandlure in traps demonstrated that certain formulations were as effective as caged males, but only for two days or less. New formulations incorporating slow-release agents, humectants, diluents, and other materials at times extended the effective period of the lures to about three weeks. Several preparations were 90 to 100 percent attractive as males in field trials. Several large-scale field tests were carried out in 1973 and 1974 in thirteen cotton-growing states and five foreign countries.[37] The boll-weevil traps were also improved considerably as the lures were being developed.

A South Mississippi Pilot Boll Weevil Eradication Experiment was started in July, 1971. This project included southern Mississippi and adjoining areas of Louisiana and Alabama. The central core was surrounded by three buffer zones. Several control measures were employed in conjunction with grandlure-baited traps and grandlure-baited trap crops. The trap crops consisted of early cotton plantings, four rows wide, on which three bait stations were side-dressed

132

with a systemic insecticide, aldicarb.[37] Use of grandlure in the trap crops improved the performance of the trap crops markedly.

In the first year grandlure-baited traps were placed around the cotton fields from mid-April until mid-July. During this period 156,580 boll weevils were captured in 5,418 traps in Zone 1 of the core area, and 132,350 in 5,979 traps in Zone 2. These traps did not control the boll-weevil population, but they were very important in indicating when other control methods should be used.

The traps were used again in the fall of 1972 to assess the results of a reproduction-diapause insecticide spray program. Only 559 weevils were trapped in Zone 1 of the core area, and 1,875 in Zone 2, indicating a successful control program. In 1973 grandlure-baited traps were again placed around all of the cotton fields in Zones 1 and 2, at the rate of one trap per acre. From April 16 to August 3, 1973, only 1,436 boll weevils were captured in the core area, and 40,173 in the buffer areas. Only twenty-eight weevils were collected in the core area from May until August. Therefore the Technical Guidance Committee for the project concluded that it is technically feasible to eliminate the boll weevil as an economic pest from the United States.[37]

A boll-weevil population-suppression program was carried out in Arkansas in 1974.[37] Grandlure-baited traps plus insecticides prevented reproduction of over-wintered weevils from June 13 to July 6. The lure-baited traps alone captured 76 percent of the weevils from planting time to the pinhead-square stage of the cotton, and 95 to 96 percent from July 6 to July 31.

The pink bollworm *(Pectinophora gossypiella)* is a very destructive insect on cotton in many places.[40] The damaging larvae settle in the squares, flowers, and young bolls and thus cannot be controlled by insecticides. Insecticides are used primarily to control the adult moths. This requires frequent

133

spraying, which is costly and has undesirable side effects. Moreover, this pest develops resistance to insecticides fairly rapidly. Thus other control measures are highly desirable.

The female pink-bollworm moth produces a sex pheromone that attracts the males from a distant point, and at high concentrations it stimulates them to copulate with her.[40] It should be possible to disrupt the mating process by permeating the air with a synthetic pheromone. The males would sense the presence of the chemical everywhere but not obtain the necessary directional odor signal to enable them to find the females. In addition, continuous exposure to the odor might cause a sensory adaptation so that they are less sensitive to the odor.

The sex attractant of the pink bollworm has been identified and named gossyplure from the genus name of cotton, *Gossypium.* It consists of the *cis, cis* and *cis, trans* isomers of 7,11-hexadecadienyl acetate[40] and is available commercially. A related chemical, hexalure (*cis*-7-hexadecenyl acetate), was discovered by empirical screening before gossyplure was isolated and identified. Although it is about 100-fold less active as an attractant than gossyplure, it was very useful in early experimentation on disruption of pink-bollworm presexual communication.

Much initial research had to be done on evaporator design and on the evaporation rates of the sex attractant in relation to the effectiveness of disruption. The degree of disruption was evaluated by determining the ability of the male moths to orient toward traps containing pheromone-releasing females, which were placed in the centers of the synthetic pheromone-permeated plots, versus male orientation to similar traps in nonpermeated areas.[40] Hexalure evaporators placed at the level of the top of the cotton foliage caused greater disruption of communication between the sexes than evaporators placed midway between the ground and the top of the foliage or those placed at ground level.

On windy nights, however, the moths move down in the plant canopy, which suggests that the evaporators would then need to be moved to lower levels.

The amount of the hexalure released per night was important in disruption of communication. In tests in southern California a greater than 90 percent disruption was achieved when the amount of hexalure released was more than 10 milligrams per hectare per 10-hour night.[40]

It is interesting that some chemicals that do not attract male pink bollworms do prevent them from orienting toward the females, when used to permeate the atmosphere. Looplure (cis-7-dodecenyl acetate), the sex pheromone of the cabbage-looper moth *(Trichoplusia ni)* is such a compound. It takes ten times as much looplure as hexalure to accomplish the same amount of disruption, and it takes about one hundred times as much hexalure as gossyplure to cause an equivalent disruption. Several other chemicals have been found to cause disruption of communication in the pink-bollworm moths, but the relative effect seems to diminish as the chemicals become increasingly different from gossyplure. Dr. H. H. Shorey and his colleagues suggest that perhaps chemicals can be found that will disrupt communication between the males and females of several insect pests that infest cotton.[40]

In the summer of 1973 six cotton fields in California were supplied weekly with hexalure evaporators, which were positioned 20 to 40 meters apart throughout the fields.[41] High-release evaporators (releasing about 200 milligrams per hectare per night) were placed in three fields, and low-release evaporators (releasing 20 milligrams per hectare per night) were placed in the other three fields. Inspections of the cotton bolls in mid-August indicated that the numbers of pink-bollworm larvae were reduced by 93 percent in the high-release fields and by 83 percent in the low-release fields, compared with untreated cotton fields in the area. Other fields that were treated four to eight times with commercial applications of the insecticide carbaryl achieved no better control of larval infestations in the cotton bolls than either the high or the low release rates of hexalure. According to Shorey and his colleagues, the results indicate that the system with the low-hexalure release rates is commercially feasible for pink-bollworm control.[40]

Because of its much greater attraction for male pink-bollworm moths, gossyplure was tested in 1974. It was continuously evaporated into the air of all of the cotton fields in the Coachella Valley of California during the growing season. Evaporators were placed 40 meters apart in the fields and level with top of the foliage canopy. They were replaced with fresh evaporators every two weeks. The release rate was 5 milligrams of gossyplure per hectare per night. Because no fields were available as controls in 1974, comparisons were made with the three previous growing seasons in the area. Through mid-August the numbers of bolls infested with larvae of the pink bollworm were comparable with the numbers observed during the three previous seasons in fields that had received conventional insecticide treatments. Moreover, there was a three- to four-week delay in the onset of larval infestations in 1974, compared with previous years. Surprisingly, the cost per hectare, even on an experimental basis, was about the same as the usual expense for insecticidal control of the pink bollworm. Shorey and his colleagues concluded that development of efficient techniques for producing tiny gossyplure evaporators, such as microcapsules, and for spreading them on the foliage will provide even better control of this pest.[40]

J. A. Klun and G. A. Junk identified four compounds in extracts of homogenates of female European corn-borer abdomen tips: tetradecyl acetate and three derivatives of that compound.[42] Two of the compounds (isomers of 11-tetradecenyl acetate) were found to be strong male attrac-

136

tants; tetradecyl acetate had no effect; and the fourth compound ([E]-9-tetradecenyl acetate) lowered numbers of male European corn borers in field and laboratory tests. The (E)-9-tetradecenyl acetate also suppressed male precopulatory reactions. Thus it really belongs in the repellent category, as a mating repellent. Therefore J. A. Klun and his colleagues suggested that a blend of this compound with the 11-tetradecenyl acetates might be useful as a disruptor of mating efficiency of the European corn borer in the field.[43]

The codling moth *(Laspeyresia pomonella)* is the most important pest of deciduous fruits.[44] It attacks apples, pears, quinces, and English walnuts. Serious crop losses are prevented only by regular spray programs that begin soon after flowering and continue until harvest. Four or more sprays are often required to give total protection in heavily infested orchards. Such heavy spraying with insecticides has seriously affected the population regulation of other insect pests of fruit trees, and several of them have become more serious pests than they were several years ago. Very promising results have been obtained in integrated programs in which traps baited with the female sex pheromone codlemone (8,10-dodecadien-1-ol) are used to determine appropriate times for spraying with reduced amounts of pesticides. H. F. Madsen and J. M. Vakenti found that they could reduce a conventional schedule of three sprays to one, or at most two, well-timed sprays per growing season and still produce high-quality apples.[45] Such a program can allow the numbers of beneficial insect predators to increase while decreasing the insecticide costs of the growers and lowering insecticide contamination of the area.

The sex attractant of the red-banded leaf roller *Argyrotaenia velutinana), cis*-11-tetradecenyl acetate, has been used for several years for male moth control in New York apple orchards.[28] One sticky trap is used per tree, and the traps are rebaited every six weeks. Fruit injury in the trapped orchards has averaged 0.6 percent, which is considered to be acceptable control.

The gypsy moth *(Porthetria dispar)* is a serious pest of forest, shade, and orchard trees in Europe and the northeastern United States.[35] The adult moths emerge sometime in July or August, depending on the local climate. The male seeks out the female, guided by the wind-carried sex attractant that she emits. After mating takes place, the female lays 300 to 800 eggs. In 1970, B. A. Bierl and his colleagues identified and synthesized the gypsy-moth sex pheromone (*cis*-7,8-epoxy-2-methyl-octadecane).[46] It has been named *disparlure* after the species name of the insect.

In 1973 a large-scale test was conducted in Massachusetts to determine the effectiveness of disparlure in disrupting the mating of gypsy moths.[35] From July 6 to July 10 airplanes were used to apply microcapsules of the attractant over a 24-square-mile area at a rate of 5 grams of disparlure per hectare. An area of equivalent size was left unsprayed as a control. In July and August counts were made of the moths captured in traps baited with disparlure or females, and the fertilization of untethered females exposed on tree trunks was determined. In the untreated area the disparlure-baited traps captured 2,193 males, whereas only sixty-three males were captured in the treated area. With the female-baited traps 1,136 males were captured in the untreated area, and only one was captured in the treated area. For two and one-half weeks after application of the disparlure mating of recovered females in the treated area remained low, while 95 to 100 percent of the females in the control area were mated. Considerable disruption of mating occurred for five weeks after treatment.

In 1974 the experiment was repeated with some modifications in the same general area of Massachusetts. Three different application rates were used in different areas: 5, 10, and 20 grams of disparlure per hectare.[35] Two applications of the 10-gram dose were made two and one-half weeks apart. As in 1973, 5 grams of disparlure per hectare had a definite effect but gave less than adequate protection. The mating of exposed females was reduced by 47 percent, compared with the control area, and both the captures of males in traps and the egg-mass counts indicated about 50-percent control of the gypsy moth, compared with the untreated area. With the higher dosages mating success was reduced by 94 to 97 percent, and male captures and egg-mass counts were greatly reduced. Thus a dose of 20 grams of disparlure per hectare (8 grams per acre) was adequate to suppress markedly the gypsy-moth population.

Many other mating-disruption experiments, mass-trapping experiments using lures, and lure-baited trap-plant experiments have been run with various degrees of success. The evidence at this point indicates that chemicals affecting insect behavior have a promising future in pest-control programs in agriculture and forestry.

CONTROL OF INSECT PESTS BY HORMONES AND ANALOGUES

Strictly defined, a hormone does not cause biochemical interactions between organisms because under natural conditions hormones are chemical regulators used in the individual that produces them. Many compounds have been found in plants, however, that affect insects and certain other invertebrates as if they were hormones produced by these organisms. Also the potential uses of insect hormones or analogues in insect-pest control ensure that discussion of this will be of value.

Ecdysones. Insects have their cuticles, or skeletons, on the outside. They must shed them periodically in order to grow larger, a process called molting. Three internal secretions, or hormones, are used by insects to regulate their growth and metamorphosis from larva to pupa to adult.[1] The molting hormone, or ecdysone, is produced in the prothoracic gland in response to an activation hormone produced in the brain.[47] Ecdysones are necessary for the resorption of the old cuticle and for deposition, hardening, and tanning of the new cuticle. The third hormone required during the larval stages is the juvenile hormone, which has to be present at each molt to prevent the insect from maturing (fig. 14). The juvenile hormone is synthesized and released from two tiny glands, the corpora allata, in the head of the insect. If these glands are removed prior to a larval molt, the production of the juvenile hormone is prevented, and the larva molts precociously to a tiny pupa or adult.[48]

The molt to the pupal stage must take place in the relative absence of the juvenile hormone. Otherwise, the insect will develop abnormally into a larval-pupal monster with a mixture of larval and pupal features, or it may molt into a giant larva that can continue feeding.[48] Most such monsters and giant larvae die shortly after or during molting. The juvenile hormone must be absent also during the transformation from pupa to adult. The treatment of a pupa with a juvenile hormone results in the formation of a creature intermediate between pupa and adult, or another pupa. The final result is a deformed insect that lives for only a few days and is unable to reproduce. The juvenile hormone must be absent from insect eggs also, or the eggs will not hatch.[1]

Therefore the treatment of insect pests with the juvenile hormone at critical stages when it should be absent or very low in amount could control the population of the pest. According to Carroll Williams, juvenile hormones are restricted to insects and have no effects on other animals or plants.[1] Thus, if these hormones could be used to control insect pests, they should not have any serious side effects.

The juvenile hormone of the cecropia moth has been

HORMONES CONTROL INSECT GROWTH
DEVELOPMENT AND REPRODUCTION

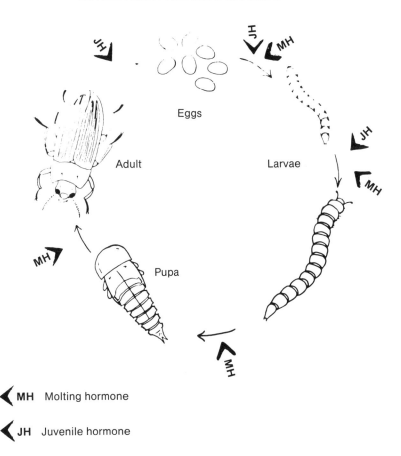

Fig. 14. Schematic representation of the endocrine control of postem-
bryonic development and reproduction in the yellow mealworm, *Tenebrio molitor.* Reproduced from W. S. Bowers, "Juvenile hormones," in
Naturally Occurring Insecticides, ed. M. Jacobson and D. G. Crosby,
courtesy of Marcel Dekker.

isolated and identified. Tests suggest that 1 gram of the pure hormone could cause the death of about one billion mealworms.[1] Originally it was thought that the juvenile hormone might be a single entity, because the crude cecropia juvenile harmone extract had morphogenetic effects on species from many insect orders.[48] Subsequent research has demonstrated that numerous compounds have juvenile hormone activity, and there is considerable variation in the response of different insects to these compounds. Slama and Williams found that paper towels placed in rearing jars of the European bug *Pyrrhocoris apterus* caused these insects to die without reaching sexual maturity. Instead of metamorphosing into normal adults, they continued to grow as larvae or molted into forms that were like adults but retained many larval characteristics. It was evident that the bugs were exposed to an unknown source of their juvenile hormone.[1] Slama and Williams extracted a factor from the towels, named it the "paper factor," and demonstrated that it acted like a juvenile hormone. Later they traced the factor back to the balsam fir tree from which the paper was made. Subsequent research has demonstrated that the paper factor occurs in fir, hemlock, yew, larch, spruce, and pine.[48]

W. S. Bowers and his group at the United States Department of Agriculture in Beltsville, Maryland, identified the paper factor from balsam-fir wood as the methyl ester of todomatuic acid, or juvabione.[1] The structure of this compound is similar to that of several other juvenile hormones that had been previously identified. It is particularly interesting, however, because the activity of the paper factor is specific for *P. apterus* and closely related insects of the Pyrrhocoridae family, which includes several serious pests of cotton. Another related compound was isolated later from fir wood and called dehydrojuvabione. It also shows activity only on insects of that family.[48] At least one other active compound is known to occur in lipid extracts of paper, and it is suspected that chemical alterations of juvabione occur during the paper-making process.

It is very probable that the juvabione and dehydrojuvabione that occur in numerous conifers are very effective in protecting those species against potential pests in the Pyrrhocoridae family. There are large numbers of other terpenoid compounds in conifers and other plants, and many of these may turn out to be analogues of the juvenile hormones of specific insect pests.[1] Readers who desire more information on the chemical structures of the numerous compounds that have been found to have juvenile-hormone activity are referred to the paper by W. S. Bowers cited in the Notes.[48]

The molting hormone, α-ecdysone, was first isolated from pupae of the silkworm *Bombyx mori.* Researchers at the Max Planck Institute in Germany isolated a very tiny amount (25 milligrams) of the pure crystalline hormone from half a ton of the pupae. This material was found to be very active in causing molting in numerous species of insects. An extremely small amount, 0.0075 μg (0.0075 of a millionth of a gram), caused 50 to 70 percent pupation in a group of fifteen larvae of the blowfly *Calliphora erythrocephala.*[3]

A second active molting hormone was isolated later from silkworm pupae and named ß-ecdysone. Both molting hormones have been isolated subsequently from many species of insects. D. H. S. Horn gives the chemical structures of α- and ß-ecdysones and forty other ecdysones from many plant and animal sources. All are steroids. α-Ecdysone has been isolated from silkworm pupae and moths, adult locusts *(Dociostaurus maroccanus),* tobacco-hornworm pupae *(Manduca sexta),* leaves of bracken fern, and rhizomes of the ferns *Polypodium vulgare, Osmunda japonica* and *O. asiatica.* It has also been synthesized.[3]

ß-Ecdysone (crustecdysone-JL1) is a crustacean and insect molting hormone. It has been isolated from seawater crayfish *(Jasus lalandei),* pupae of the oak silk moth *(Anthe-*

raea pernyi), silkworm pupae, tobacco hornworms, the roots of *Achyranthes fauriei,* wood and bark of *Podocarpus elatus,* leaves of the agnus castus, or chaste tree *(Vitex megapotamica),* rhizomes of the fern *Polypodium vulgare,* mulberry leaves, roots of *Bosea yervamora,* leaves of yew, roots of *Achyranthes obtusifolia,* leaves of bracken fern, and the whole plants of *A. obtusifolia, A. longifolia, Matteuccia struthiopteris,* ajuga *(Ajuga decumbens, A. incisa),* wake-robin *(Trillium smallii, T. tschonskii), Stachyurus praecox, Polypodium japonicum, Pleopeltis thunbergiana,* and *Neocheiropteris ensata.*[3] It has also been synthesized. Of the forty additional ecdysones for which structural formulae are given by D. H. S. Horn, a few have been isolated only from animals, and a few have been obtained only by synthesis, but most have been found only in plants. Horn also lists 116 species of plants that have been found to be active in insect tests for ecdysones.

After summarizing the tremendous amount of work done on ecdysones, Horn states that it is remarkable that compounds with both molting and juvenile hormone activity have been found in plants.[3] He suggests, "These compounds may provide, or have provided, some measure of protection against insect attack." As insect growth and development is controlled by a maximum concentration on the order of one part per million, plants in which ecdysones constitute as much as 0.1 percent of the dry weight must present a considerable biochemical problem to insects attempting to eat them. Apparently many insects that feed on a wide variety of plants have developed detoxification mechanisms against these plant-produced ecdysones.

The ecdysones and analogues are generally not as active as the most active juvenile hormones, but they still are able to disrupt insect development and reproduction when applied directly to the insect or added to the diet in microgram quantities.[49] They are more expensive to produce synthetically, because they must have a steroid nucleus to be active. They also penetrate the cuticle less readily than the smaller juvenile-hormone molecules. A major advantage of the ecdysones and analogues is that they are able to disrupt development during any larval or nymphal stage. These compounds also act as female chemosterilants in insects, because they inhibit ovarian development and egg production. They also reduce egg hatch and the viability of the progeny.

D. H. S. Horn[3] and W. E. Robbins[49] have suggested that a promising approach to insect control by means of hormones is to develop and use compounds that interfere with the biosynthesis and metabolism of the ecdysones in the insects. Only very limited field tests of the juvenile hormones and ecdysones, or their analogues, have been made. Moreover, most of the research has been done on organisms that are not particularly important in agriculture. In 1982 it was difficult to find references to the actual use of these compounds in insect-control programs in crops. They did not appear to be as practical as the behavior-affecting chemicals.

CHAPTER 7

Chemical Interactions Involving Crop Plants and Animals Other than Insects and Nematodes (Including Pheromones)

THERE are two basic reasons to stress the insects and nematodes in discussing chemical interactions between animals and crop plants: (1) insects are the chief consumers of plants in the world, and (2) much more is known about chemical interactions between crop plants and nematodes than about chemical interactions between crop plants and other animals. Yet, though our knowledge of plant interactions with other animals is fragmentary, a brief discussion will indicate the potential for future developments. Considerable damage is done to crop plants in some parts of the world by molluscs (certain snails and slugs), mites, rabbits and hares, rodents, birds, and some of the hoofed animals. Obviously, other animals can be harmful also, but generally they are less so. Because of the sparseness of information some examples of interactions with noncrop plants will be given.

Most people are less familiar with the characteristics of mites than with the other animals listed above. They are tiny relatives of the spiders and scorpions and are placed in the class Arachnida. Along with other differences, these organisms have four pairs of legs rather than the three pairs that the insects have. Possibly the mites exceed in numbers of individuals all of the other arthropods on the planet.[1] Almost any handful of soil contains large numbers of them. The chigger, for example, is a newly hatched mite that bores through the human skin and injects a poison that causes an itching welt to form. Many other mites also feed on warm-blooded animals as do their close relatives the ticks.

ATTRACTANTS AND REPELLENTS IN PLANTS OR STORED FOOD PRODUCTS

Unlike the scientific literature on insects, there is little published evidence of the production in plants of chemical compounds that attract or repel other animals. It seems highly probable that the mites and at least some other animals locate their host plants because of volatile attractants produced by the plants. It also seems likely that animal pests locate many stored plant materials and other stored food products because of attractants produced by the stored materials, though little research has been done in this area. Different varieties of strawberries differ in resistance to the two-spotted mite *Tetranychus urticae,* and the differences are hereditary.[2] J. G. Rodriguez and his colleagues tested water extracts of three genetically different clones of strawberries for attraction or repulsion of *T. urticae* and the strawberry spider mite, *T. turkestani.*[3] The clones were Ky 22-61-9, Ky 17-61-15, and Citation, of which the first is the most resistant and the last the most susceptible to the mite. Both species of mites were attracted to the odor of the Citation extract in all of the dilutions tested. In a test in which two different extracts were compared at the same concentration, both mite species were attracted to the odor of the Citation extract over that of either of the two other clones.

The odor of the most concentrated extract of Ky 17-61-15

was repellent to the two-spotted mite, but not to the strawberry spider mite. No other extract or dilution was repellent to either mite. Subsequently, Z. T. Dabrowski and J. G. Rodriguez tested the responses of the two mites against eleven essential oil components from Citation strawberry leaves.[4] One of the components, *trans*-2-hexen-1-ol, was an attactant or repellent to both mites, depending on the concentration. Five other components elicited the same kind of concentration-dependent response from the strawberry spider mite, but not from the two-spotted mite. Five of the components acted as repellents to the two-spotted mite at all of the concentrations tested, and one of them acted as a repellent to the strawberry spider mite at all of the concentrations tested. Some had no effect alone, but each had effects when combined in a 1:1 mixture with another component.

The resistance of a strawberry clone to mite predation is known to change during the season. It seems likely that a preferred host can change to a nonpreferred host in a relatively short period of time because of biochemical changes within it.

In Japan many important food products, such as cheese, spices, bean paste, dried fish, and powdered milk, are infested by mites.[5] This causes serious losses and sanitation problems. The cheese mite *Tyrophagus putrescentiae* is a serious pest in Japan. Preliminary tests showed that it is attracted to cheddar cheese by odor. Subsequent research demonstrated that the neutral fraction of the steam distillate of cheddar cheese contained an attractive component. This component was separated by gas chromatography but was not identified.

Many species of plants are pollinated by birds and bats. The birds must locate appropriate flowers by sight, because they have scarcely any sense of smell and bird-pollinated flowers typically lack an odor. Bats, on the other hand, have a good sense of smell, and at night bat-pollinated flowers typically have a strong odor reminiscent of butyric acid. The odor of the flowers is similar to the odor of the bats themselves; it is thought to have some social function in their aggregation and may have a stimulatory effect. Unfortunately, there is no solid experimental evidence that there are specific bat attractants in bat-pollinated flowers. Such a chemical investigation would be highly desirable.[6] Although this has little to do with crop plants, some cultivated tropical fruit trees are bat-pollinated.

FEEDING DETERRENTS PRODUCED BY PLANTS

J. G. Rodriguez and his colleagues studied the rate of ingestion and total uptake by mites of water extracts of leaves of the three strawberry clones discussed above, Ky 22-61-9, Ky 17-61-15, and Citation.[3] They found that the rate of ingestion and the total uptake of the Citation extract by the two-spotted mite far exceeded its uptake of the extracts of the mite-resistant clones, Ky 22-61-9 and Ky 17-61-15. This suggested the presence either of feeding deterrents in the extracts of the resistant strains or feeding stimulants in the susceptible strawberry strain.

Z. T. Dabrowski and his colleagues extended this research and found that another mite, *Panonychus ulmi,* fed well only on the leaves of selected plant species when many types of leaves were offered as food sources. The reaction of the mites to the undesired leaves suggested to the researchers that those leaves contained chemicals that acted as feeding deterrents. They tested many compounds from strawberry foliage against feeding by the two-spotted mite and found that several free amino acids were feeding deterrents in a 0.1 percent solution.[7] They found also that very dilute solutions (10^{-4} to 10^{-6} M) of most phenolic compounds tested were feeding deterrents to the two-spotted mite.

149

Thus evidence is slowly accumulating that the location and selection of host plants by mites depends strongly on the chemical compounds produced by the plants, just as has been demonstrated for many insects.

Wild populations of bird's-foot trefoil *(Lotus corniculatus)* include some individuals that produce a cyanogenic glucoside and some that do not; and this character is genetically controlled. While sampling wild populations of bird's-foot trefoil for the frequency of the cyanogenic plants, David Jones noticed that generally only the acyanogenic plants (those containing no cyanogenic glucosides) were grazed by animals.[8] The only animals that he observed actually feeding on the acyanogenic plants were insect larvae and slugs. In a series of experiments cyanogenic and acyanogenic bird's-foot trefoil plants were made available to slugs of the species *Arion ater.* The plants that contained no cyanogenic glucoside or the least cyanogenic glucoside were always eaten the most. In less extensive tests good evidence was obtained that the vole *Microtus agrestis,* two snails *(Arianta arbustorum* and *Helix agrestis),* and the slug *Agriolimax reticulatus* also show this selective preference for acyanogenic bird's-foot trefoil plants.

White clover *(Trifolium repens)* is also polymorphic for the presence or absence of cyanogenic glucosides. T. J. Crawford-Sidebotham decided to do a comprehensive series of laboratory tests on the feeding of several snails and slugs on the two forms of both white clover and bird's-foot trefoil.[9] He used seven snail and six slug species in the experiments. The results support the hypothesis that snails and slugs selectively eat acyanogenic plants and therefore that the cyanogenic character acts as a defensive mechanism against them. The various species of slugs and snails differed, however, in the degree to which they exhibited differential eating, and the differential selection seemed to be more clear-cut in bird's-foot trefoil than in white clover.

W. M. Ellis and his colleagues found a differential distribution of the cyanogenic and acyanogenic forms of bird's-foot trefoil around the bay at Porthdafarch on Holy Island, England.[10] The acyanogenic form was less frequent as one moved inland from the sea cliff, and this pattern of distribution was stable over a sixteen-year period. Moreover, the population of snails and slugs was low near the sea cliff and increased with distance from the cliffs. After studying several ecological factors for sixteen years, the researchers concluded that the distribution of the cyanogenic form of the plant is directly related to the distribution and density of the slugs and snails that graze selectively on the acyanogenic form.

Most populations of white clover in Europe are mixtures of cyanogenic and acyanogenic plants. J. P. A. and W. J. Angseesing decided to determine if differential eating of the two forms occurs under field conditions, as had been demonstrated previously under laboratory conditions.[11] The study was carried out on a clover-rich lawn in Cheltenham, England. Leaves were picked at random and typed according to the quantity of each leaflet that had been eaten by herbivores, after which the concentration of cyanogenic glucoside in each leaflet was determined. The results clearly indicated that acyanogenic clover plants were more heavily eaten than the cyanogenic ones. The principle grazers were found to be the slugs *Arion ater* and *Agriolimax reticulatus.*

W. A. Dement and H. A. Mooney observed that fruits of the chaparral shrub *Heteromeles arbutifolia* have a long maturation period characterized by low levels of predation by animals.[12] On maturation, however, the fruits are removed rapidly by birds. Chemical analyses of the various plant parts at different developmental stages demonstrated that secondary plant products seem to be important in this differential predation. Immature fruits have extremely high tannin levels plus cyanogenic glucosides in the pulp. On maturation the tannin levels decline, and the cyanogenic glucosides are shifted from the fruit pulp to the seeds. The leaves have high concentrations of both tannins and cyanogenic glucosides at the time of their initiation, which probably decreases animal feeding in the early stages of growth.

Jojoba *(Simmondsia chinensis)* is an evergreen shrub distributed throughout the southwestern United States and northwestern Mexico. The pistillate, or female, plant produces large seeds that are a considerable food resource to animals.[13] M. L. Rosenzweig and J. Winakur captured the pocket mouse *Perognathus baileyi* on all of the research plots on which jojoba grew, but on other plots another pocket mouse was captured.[14] They suggested that there may be a direct connection between the distributions of *P. baileyi* and jojoba.

Since jojoba seeds are known to contain appreciable amounts of cyanogenic glucosides, Wade Sherbrooke decided to place four species of desert rodents, including *P. baileyi,* on a diet of the seeds.[13] *P. baileyi* lived on a diet of the seeds or meal made from the seeds for several weeks. The other three rodent species, which included two other species of *Perognathus,* refused to eat jojoba seeds and lost weight rapidly. Sherbrooke suggested that *P. baileyi* apparently has a detoxification mechanism not possessed by the other species. He suggested also that the cyanogenic glucosides in the jojoba seeds may function as a defense in seed predation.

Bracken ferns have been found to be polymorphic for prunasin, a cyanogenic glucoside. G. A. Cooper-Driver and T. Swain correlated herbivore predation with changes in content of prunasin in Richmond Park, Surrey (England).[15] They found that in most populations 96 percent of the individuals were cyanogenic, but in a few populations 98 percent of the individuals were acyanogenic. In the latter the bracken fronds were heavily grazed by deer and sheep, while in the former stands only fronds from the acyanogenic plants were grazed.

Several species of the composite family, which is characterized by high concentrations of sesquiterpene lactones, remain virtually untouched by cattle in heavily grazed pastures in eastern North America. Some of the species belong to the ironweed genus, *Vernonia,* and contain glaucolide-A, a sesquiterpene lactone that is a feeding deterrent to several insects. W. C. Burnett and his colleagues decided to determine whether that compound is also a feeding deterrent to mammals.[16] They observed that wild rabbits in the Botany Transplant Garden at the University of Georgia fed heavily on *Vernonia flaccidifolia,* which lacks glaucolide-A, while sampling and rejecting *V. gigantea* and *V. glauca,* which contain this compound. When caged wild eastern cottontail rabbits *(Sylvilagus floridanus)* were offered *V. flaccidifolia* and *V. gigantea,* they consistently fed heavily on the former but only sampled the latter. In another test they preferred natural *V. flaccidifolia* to *V. flaccidifolia* coated with glaucolide-A.

The whitetail deer *(Odocoileus virginianus)* occurs naturally in the same region with the three species of ironweed discussed above. Burnett and his colleagues tested its feeding preference by offering a choice between *V. gigantea* and *V. flaccidifolia.* The deer would smell, lick, and sample both plants. Then they consistently chose to feed on the plant lacking glaucolide-A. When they were given a choice between natural *V. flaccidifolia* and *V. flaccidifolia* coated with glaucolide-A, they always showed a preference for the noncoated plant. Most ironweed plants in nature are protected by the presence of this bitter compound, and an ironweed plant without it may be avoided because of its similarity in appearance to the others. It seems likely that a great many other sesquiterpene lactones—perhaps all—serve as

153

feeding deterrents in a great many species of plants. There are some 540 known sesquiterpene lactones in the composite family alone.[16]

Douglas fir *(Pseudotsuga menziesii)* is generally highly susceptible to animal damage and frequently needs protection in young plantations. The black-tailed deer *(Odocoileus himionus columbianus)* and the snowshoe hare *(Lepus americanus)* are considered serious obstacles to reforestation in the Pacific Northwest. Therefore Edward Dimock II and his colleagues decided to determine if different genetic types of Douglas fir would affect feeding selection of captive snowshoe hares and black-tailed deer in pen tests.[17] They found that the preferred genotypes were, respectively, 64 and 178 percent more attractive to the deer and hares than those least preferred. The order and the magnitude of damage resistance predicted for offsprings, based on the animals' preferences among clones, closely conformed to the resistance traits indicated for the parents. The results of subsequent field tests with wild hares agreed closely with the pen tests.

M. A. Radwan and G. L. Crouch attempted to identify some of the chemical factors in Douglas fir that cause the differential responses of black-tailed deer in feeding tests.[18] Douglas-fir foliage was obtained during the dormant season from eight genetic strains and analyzed for chlorogenic acid (a phenolic acid) and for essential oils. The foliage from the same trees was ranked according to the browsing preference of black-tailed deer. The researchers found that the preference order correlated significantly with the chlorogenic-acid concentration in the foliage. Specifically, greater susceptibility to deer browsing was associated with higher concentrations of chlorogenic acid—an indication that chlorogenic acid is a feeding stimulant to deer rather than a feeding deterrent. The genetic strains varied significantly in the yields and composition of essential oils also, but the differences were not related to browsing preference. These results certainly indicate the feasibility of selecting and breeding forest trees for resistance to browsing.

Reed canary grass *(Phalaris arundinacea)* is a commonly planted forage and hay crop in many parts of the world. There have been numerous reports of marked differences in its palatability to domestic animals. R. Roe and B. E. Mottershead decided to run feeding preference tests with five strains of the grass, using Merino sheep as test animals.[19] They tested two strains of the grass from the United States, two from West Germany, and one from Portugal. Three of the strains proved to be palatable to the sheep, and two were very unpalatable. One of the latter came from the United States, and the other from Portugal. Subsequently, Martin Williams and his colleagues analyzed palatable and unpalatable genotypes of reed canary grass for types and amounts of alkaloids.[20] Four genotypes were chosen, and palatability ratings were established by grazing tests with sheep. Six alkaloids were identified, and three additional compounds were isolated but not identified. The alkaloid content of the palatable and unpalatable clones averaged 0.26 and 0.78 percent of dry weight, respectively. The amount of one of the alkaloids (5-methoxy-N,N-dimethyltryptamine) in the unpalatable clones averaged eighteen times the amount found in the palatable clones. Thus some or all of the alkaloids in reed canary grass are feeding deterrents to at least some mammals.

Peter Atsatt and Dennis O'Dowd state that certain forage grasses "derive considerable protection from cattle when associated with the noxious buttercup, *Ranunculus bulbosus.*"[21] They point out that this species contains the lactone ranunculin, which is a powerful irritant of skin and mucous membranes. In one study cattle grazing decreased significantly where the buttercups were more dense. The same authors state that the poisonous burroweed *(Haplopappus tenuisectus)* conveys a considerable degree of protection from grazing to neighboring grasses in southern Arizona.

155

Many species of buttercup contain ranunculin, and several species contain much higher concentrations than *R. bulbosus*. John Kingsbury's *Poisonous Plants of the United States and Canada* discusses many of the vascular plants found in North America that can be toxic to domestic animals,[22] including numerous crop plants. Since various animals know the plants to be distasteful, the toxic compounds often act as feeding deterrents.

PLANT-PRODUCED COMPOUNDS WITH DELAYED EFFECTS ON ANIMALS

Certain fatty acids have been shown to inhibit the growth of algae[23] and various insects.[24] J. G. Rodriguez decided therefore to determine the effects of some twenty-five fatty acids and their chemical derivatives on the growth and egg laying of the cheese mites *Tyrophagus putrescentiae* and *Caloglyphus berlesei.*[24] Various percentages of fatty acids were added to the diet of newly emerged larvae. Then survival rates, development, and numbers of eggs laid were observed (figs. 15 and 16). Since most of the detailed work was done with *T. putrescentiae,* a summary of the results with this organism is given below. The results with *C. berlesei* were similar in the limited comparative tests that were run.

Fatty Acids. The short-chain fatty acids—such as propionic, butyric, caproic, caprylic, and capric acids—were especially effective in inhibiting development and survival of *T. putrescentiae*. Very few eggs were produced by these mites when cultured in a food preparation containing 0.5 percent capric acid, and no eggs were produced at a 1.0 percent concentration. All of the other fatty acids tested required a 2.0 percent concentration to produce this level of inhibition. Copper (cupric) salts of propionic, caproic, caprylic, and capric acids were also extremely effective as growth inhibitors.

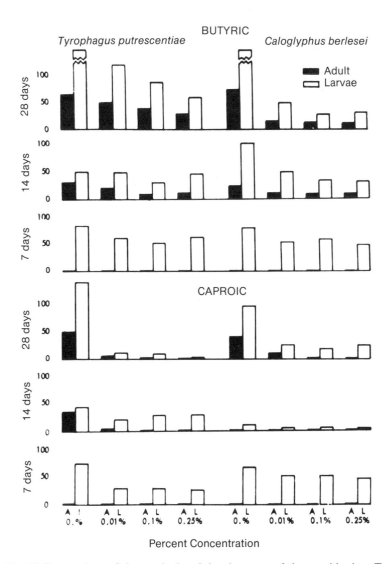

Fig. 15. Comparison of the survival and development of the acarid mites *Tyrophagus putrescentiae* and *Caloglyphus berlesei* when two fatty acids, butyric and caproic, were added to a chemically defined diet in concentrations of 0.01 to 0.25 percent. The data represent the mean of two experiments, each having treatments composed of five tubes inoculated with 20 larvae per tube. Reproduced from J. G. Rodriguez, ed., *Insect and Mite Nutrition*, pp. 637-50, courtesy of Elsevier Biomedical Press B.V., Amsterdam.

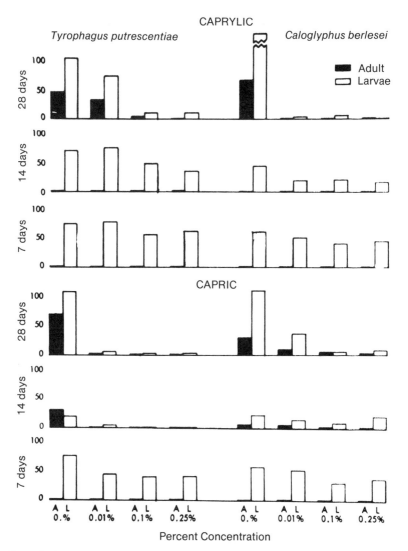

Fig. 16. Comparison of the survival and development of the acarid mites *T. putrescentiae* and *C. berlesei* when two fatty acids, caprylic and capric, were added to a chemically defined diet in concentrations of 0.01 to 0.25 percent. The data represent the mean of two experiments, each having treatments composed of five tubes inoculated with 20 larvae per tube. Reproduced from J. G. Rodriguez, ed., *Insect and Mite Nutrition,* pp. 637-50, courtesy of Elsevier Biomedical Press B.V., Amsterdam.

Methyl esters of caproic, caprylic, capric, lauric, myristic, and oleic acids were very inhibitory to *T. putrescentiae,* and they were effective at a concentration of 0.5 percent except for the esters of caproic and caprylic acids, which required 1.0 percent concentration to be effective.

The evidence suggests that adding an inhibitory quantity of a free fatty acid or its derivative to a processed food would protect the food from mites.[24] It appears that the concentrations of certain free fatty acids in stored food products from plants could also be important in protecting the products from mites. The concentrations could probably be controlled in many plants by appropriate selection and breeding programs.

Tannins. Tannins are a group of phenolic compounds that are widely distributed in plants and prevalent in many vegetable foodstuffs.[25] Dietary tannins from natural foodstuffs have been shown to depress the growth of chickens[26,27] and rats.[28] P. Vohra and his coworkers reported that tannic acid depressed the growth of chickens, even in concentrations as low as 0.5 percent.[27] Other tannins were less toxic. M. A. Joslyn and Z. Glick found that tannic acid was more inhibitory to the growth of rats than several other tannins at a 1 percent concentration in the diet.[28] All of the tannins significantly inhibited rat growth when present in a 5 percent concentration in the diet. Moreover, two phenolics, gallic acid and catechin, significantly inhibited the growth of rats in relatively low concentrations in the diet.

Margaret Peaslee and Frank Einhellig found that albino male mice with an initial body weight of 7 to 11 grams were inhibited in growth by an 8 percent concentration of tannic acid in their diet.[29] The growth rate of similar mice with initial body weights of 17 to 18 grams was initially depressed by a diet containing tannin, but their weights were not different from those of the controls at the time of autopsy.

In a subsequent study by these researchers female mice of the same albino strain, when fed on a similar diet containing tannin, required more time than control mice to become pregnant.[30] They also produced smaller litters, which had slower growth rates and lower final body weights.

Tannic acid has been implicated as a cancer-causing agent in some instances.[31] J. F. Morton[32] claims that the British put milk in their tea to condense the tannins and thus prevent esophageal cancer. Certainly the milk would condense the tannins, because the proteins in the milk would react readily with them, just as the proteins in hides do during the tanning process.

Indirect Effects of Plant-Produced Chemicals on Ruminants

The chemical makeup of forage and crop plants often has pronounced effects on the digestibility and nutritional value of the plants to ruminants because of the effects of certain chemicals on microorganisms in the rumen of the animals. Robert Hungate gives a very thorough discussion of all aspects of the rumen and its microorganisms for those readers desiring more background information.[33] W. W. G. Smart and his colleagues reported that polyphenolic compounds in sericea lespedeza leaves inhibit the activity of the cellulase in the rumen, which is responsible for the digestion of cellulose.[34] The enzyme cellulase is, of course, produced by microorganisms in the rumen. Essential oils in sagebrush *(Artemisia tridentata)* have been shown to inhibit the growth of several species of bacteria.[35] When essential oils from this sagebrush are added to alfalfa hay in the rumen fluids of deer, sheep, or cattle in artificial rumen systems, the production of short-chain fatty acids is decreased.

Some samples of orchard-grass hay *(Dactylis glomerata)* cause poor growth, stiffness, and even death when fed to lambs.[36] Extracts of such samples were found to contain

160

chemicals that markedly inhibit cellulose digestion in artificial rumen systems.

The tannin content of sorghum grain and forage definitely affects the digestibility of these materials. Undoubtedly this is because of the tannins' pronounced antimicrobial action. Since tannins unite easily with proteins, and enzymes are, of course, proteins, tannins are also directly effective in inactivating the enzymes in the rumen.[37,38]

There is little doubt that a great many other secondary plant products have detrimental effects on the microorganisms and enzymes in rumen and thus on the growth, development, and reproduction of the ruminants. Research in this area will increase in the future because of the great economic importance of the ruminants as a food source.

PHEROMONES

Most animals have to communicate with others of their own species if they are to survive. This necessity is obvious for animals, such as honeybees, that live in complex societies. For those that generally live alone, it is less obvious, but there are certain times when they also must communicate in some way. In bisexual animals some form of communication is required to bring the two sexes together and stimulate and guide the process of copulation or the union of gametes.[39] Of the several types of stimuli and sense organs that may be used, the basic ones are chemical, affecting smell and taste; mechanical, based on feel and sound; and radiational, based on visual or light perception.

Chemical communication among animals of the same species occurs throughout the animal kingdom. It appears to be the chief means of communication in most types of animals,[39] though the birds and the higher primates seem to be two major exceptions to the general rule. It is unfortunate that our knowledge of animal communication by pheromones is comparatively sketchy. There is considerable general information, but specific facts on pheromone production in animals other than insects is fragmentary, and relatively few specific pheromones have been chemically identified.[40-42] The potential for future research in this broad field is tremendous. Unquestionably new knowledge will result in important practical applications in animal pest control, in agriculture, and elsewhere.

The phenomenon of territoriality in animals has been recognized since the time of Aristotle. Although it has long been associated with defense and aggression, more and more emphasis is being placed on scent marking. According to R. Mykytowycz, all animals do as men do and reserve separate sections of their home ranges for different purposes, such as sleeping, resting, sheltering, eating, food storage, drinking, defecation, urination, wallowing, and the disposition of young during infancy.[40] In all those spatial requirements odor marking is characteristically used and plays an important role in defining the different areas.

Odor Marking

The space requirements of mammals are matched by their ability to mark differentially the various kinds of space that they require. There is evidence that the size and activity of odor-producing glands change in relation to the individual's sex, reproductive state, age, social status, and behavioral state. According to R. Mykytowycz, there are variations in the ability of animals to perceive odors also.[40] Moreover, age and social status affect the responses of the individual to odor signals. For example, though territoriality is strictly observed by adults of the wild European rabbit *(Oryctolagus cuniculus)* during the height of the breeding season, some juveniles will disregard territorial boundaries and move into new group territories.

The odors used to mark space come from various sources: urine, feces, saliva, and special odor-producing skin glands. In the rabbit two externally secreting skin glands function in space marking. The secretion of the submandibular gland is applied to objects as the rabbit touches its chin to them, and the scent from the anal gland coats the fecal pellets as they pass through the end of the rectum (fig. 17). The characteristic odor of rabbit feces comes from the latter gland, and dunghills in the home range of the rabbit are important objects in the territorial behavior of the species.[40]

Rabbits. The secretions from the skin gland on the chin of the rabbit contain proteins and carbohydrates, and those from the anal gland contain proteins and lipids.[40] The free lipid with the rabbit odor may be an ester of a fatty acid. There appear to be variations in the composition of the secretions of each gland between the sexes and among individuals.

The odors associated with the space occupied by various animals can be detected even by man. It is not difficult to detect rat- and mouse-infested areas or to differentiate a horse stable from a dairy shed by odor alone.[40] Such characteristic species odors are due both to spontaneously deposited metabolic products and to space-marking activities by the animals; they are generalized combinations of odors. Many animals can detect a strange population of the same species because of its characteristic odor.

Mykytowycz points out that the behavior of a rabbit entering the home range of a strange colony suggests that it is reacting to odor.[40] It changes its posture, moves cautiously, and continually sniffs at its surrounding. It also stops eating and marking and displays a submissive posture to the permanent residents of the area. Adult female rabbits often severely attack baby rabbits (kittens) born within a strange colony, and blindfolding of the adults demonstrates that such

Fig. 17. Rabbits use their chin glands to scent-mark objects, such as branches, in their territories. Reproduced from H. H. Shorey, *Animal Communication by Pheromones,* courtesy of Academic Press.

behavior is caused by detection of a strange odor. It is important to a young rabbit's survival that it recognize the characteristic odor of its own group early in life.

Rabbit litters are born in specially dug chambers that have been previously lined with grass and fur that the mother plucks from her own belly, chest, and flanks.[40] The mother also deposits a few fecal pellets in the nest; the kittens apparently learn early to recognize the anal odor of the mother and to pay more attention to it than to other rabbit odors. Except at feeding time, the entrance to the chamber is blocked with a soil plug, on which the mother deposits a few fecal pellets and some urine. Generally other members of the group will not enter the chambers unless population pressures become too great. Under the stress of high populations other females frequently kill the young.

Since rabbits are often serious predators of crop plants and pastures, an increased knowledge of their scent-marking pheromones might enable us to use formulations of them to disrupt completely the behavior of rabbits in the crop or pasture area, achieving destruction of the young and perhaps the complete withdrawal of rabbits from the area. This would appear to be a possibility without serious side effects, such as the elimination of other desirable animals or pollution of the area with persistent chemicals.

The greatest proportion of the small mammals live either beneath the surface of the ground or on the surface where denseness of vegetation or other barriers generally preclude visual or acoustic communication except for close-range behavior patterns. In such habitats it seems likely that natural selection has favored the survival and spread of species with well-developed scent-producing organs. Rodents and other small mammals are richly endowed with odor-producing organs of a specific nature, according to D. Michael Stoddart.[41] The principal sources of their odors are glands and gland complexes located in the skin.

Mice. Studies of the aggressive interactions in mice have demonstrated that mice definitely produce a factor, or factors, which induces fear of attack in other mice.[41] This substance seems to be taken up by the surroundings and remains active for seven or eight hours. Such a fear-inducing pheromone could be used as a mice repellent in gardens or other crops if properly identified and synthesized. Other compounds could be combined with it to prevent its breakdown or rapid evaporation and thus prolong its effectiveness, as has been done with various insect pheromones and repellents.

J. M. Bowers and B. K. Alexander have found considerable specificity in the pheromones produced by mice.[43] Using male and female mice of the species *Mus musculus* (strain C57B1), they tested the animals' ability to discriminate between various target mice on the basis of odor alone. They found that the mice were able to differentiate between two male mice of the same strain, purely on the basis of odor. Moreover, they were able to discriminate by olfactory cues between two different species, *Mus musculus* (strain C3H) and *Peromyscus maniculatus,* and between males and females of a given species.

It is noteworthy that the presence of sex attractants has not been clearly demonstrated for mice, particularly the female-produced attractants of males. The opposite attraction is better documented. Female mice become more active in the presence of odor from males, and females prefer urine collected from intact males to urine from castrates. Lipids from the preputial gland of the male have been implicated as attractants, specifically some of the free fatty acids.[44] Pregnant females are not attracted to the male's preputial odor.

Examples of olfactory attraction to a receptive female have been reported for dogs, deer mice, deer, and rats.[44] Such an olfactory sex attraction has been observed to work both ways for some animals, as in receptive ewes, which will successfully seek out a ram without the use of visual cues.

Rats. J. B. Calhoun observed that the receptive female Norway rat living under seminatural conditions ranges farther than normal from its burrow and leaves scent markings on the ground and other objects.[45] These scent marks are examined by males and guide them to the burrow of the receptive female. Subsequently W. J. Carr and his coworkers studied the response of rats of the Long-Evans strain to the odor of individuals of the opposite sex under controlled conditions.[46] They found that experienced males preferred the odor of a receptive female over that of a nonreceptive female, but naive males and castrates showed no preference. In another experiment they found that naive receptive females showed the same preference for the normal male odor as experienced receptive females did, whereas naive nonreceptive females showed no preference. These results indicate that sex odors play a role in the reproductive behavior of rodents by causing the sexes to be mutually attractive at an appropriate receptive period in the female.

In the white rat and mouse the preputial glands release an attractant odor affecting animals of the opposite sex. Production of the odor is dependent on the hormonal condition of the donor. Maria Stacewicz-Sapuntzakis and Anthony Gawienowski identified five very volatile aliphatic acetates from the preputial gland of male Sprague-Dawley rats.[47] They tested the attraction of these compounds for male and female rats, along with many other related compounds, most of which had previously been identified from the preputial gland of the mouse. Ten of the saturated and four of the unsaturated acetates elicited highly significant responses from the female rats, but none of the males displayed much interest. In fact, the males seemed to avoid some of the acetates. Among the acetates that these researchers found, propyl, pentyl, and decyl acetates were highly significant as attractant odors for female rats, while the male rats were indifferent to those odors. Stacewicz-Sapuntzakis and Gawienowski concluded that aliphatic acetates may contribute to the sex-specific odor of rat preputial gland secretions. Their conclusion was based on the difference in acetate content between male and female preputial glands and the sex-dependent receptivity of the animals.

Gawienowski and his coworkers decided to test fifteen compounds, including some acetates, against wild rodents.[48] These compounds had previously been shown to attract rats in laboratory tests. Some compounds attracted just males, some attracted just females, and some attracted both sexes. They designed an apparatus that would automatically monitor the response of the rodents to the compounds, and they placed it in a corner of a shelter where horse food was stored. The shelter had been bothered a lot by rats and mice consuming the food. The concentrations of the test compounds were at the lowest levels detectable by laboratory personnel. Many of the compounds proved attractive to wild rodents. Particularly attractive were arachidonyl acetate, pentyl acetate, dimethyl sulfite, hexenal, decanoic acid, 2-acetylpyridine, acetal, 4-methyl-3-pentene-2-one, and decanol. In this experiment the sex and species of the visiting rodents were not determined. This would need to be done before the attractants could be tested in rodent control programs.

It appears that in most mammals smell is involved at the different stages of reproduction under natural conditions.[49] Smell is involved in the acquisition of breeding space, social status, premating behavior, mating, the relationship between mother and children, and imprinting to the species and social unit. Under domestication there has been selection against many forms of behavior that depend on communication by odors. Nevertheless, domestic animals still retain their abilities to produce and detect odors. The sources of odors in animals include vaginal discharge, urine, saliva, feces, skin glands, seminal fluid, and embryonic fluid. All have been demonstrated to contribute to the communication related to reproduction in domestic animals. In fact, the earliest experimental reports on the role of smell in the reproduction of mammals came from observations of domestic species.

The stimulating effect of vaginal discharge on male sexuality has been demonstrated in some species. In rhesus monkeys this discharge contains mixtures of volatile fatty acids that have sex-attractant properties. In many domestic animals the raising and rapid waving of the tail, presentation of the hindquarters, and exposure of the clitoris encourage nosing by the male, which is often followed by massaging and licking of the mucus secretion. When the rear sections of nonestrous ewes are smeared with the vaginal discharge from estrous ewes, males are attracted to them and sometimes attempt copulation.[49]

The concentrations of the steroids present in urine reflect clearly the reproductive status of the individual mammal. Some of the steroids have characteristic odors that are detectable even by the human nose.[49] Regular sampling of the female's urine is a daily routine of the males of many species, and they can detect the estrous long before the characteristic behavior or changes in the external genital organs are evident.[50] When approached by a male, females usually facilitate examination by producing small quantities of urine.

Deliberate urination during reproductive activities is characteristic of males of many species as well as females.[49] The males use their urine not so much to signal their reproductive status as to indicate social dominance and control of the space in which the interaction takes place. Odor increases the influence of an individual and often repels competitors without fighting. The male reindeer digs with his forefeet and urinates on the soil, after which he rubs his nose in the urine patches. Females rub their noses in the patches also. Billy goats and some other male animals urinate on themselves during the breeding season. There is some evidence that the presence of a boar will accelerate puberty in female pigs in response to a pheromone in the male urine.

Just as fecal odors are very important in communication among certain free-living animals, there is some evidence that this is true in some domestic species also. Bulls spend more time, for example, examining the feces of the estrous than the nonestrous cows.[51] C. A. Donovan found that the application of feces of estrous cows to the tails of the nonestrous cows increased the attractiveness of the latter to males. Bulls of some breeds of cattle frequently pass small amounts of feces during mating, accompanied by pumping of the tail.[49] The tail pumping suggests that massaging of the anal glands may be involved.

Most sex-attractant odors come from the ano-genital region, but a few come from the head region.[49] The saliva is one source of odor in the head region. A chemical ($5a$-androst-16-en-3-one) with a urinelike smell has been identified as the chief component in saliva of the domestic pig. The submaxillary gland acts as the reservoir for this chemical, and this gland is more strongly developed in the male than in the female. The head-to-head position appears to be the most frequent one when a boar meets a receptive female. Salivation by the male occurs, and this seems to cause the sow to go into the mating stance. It has been demonstrated that exposure of the estrous sow to the odor alone generally elicits the mating stance. Aerosols containing steroid compounds found in the saliva of male pigs are now used to detect estrous in sows. If a sow is estrous, the spray releases the mating stance when blown into her face.

The secretion from the boar's preputial gland and his seminal fluid can elicit the mating stance in the estrous sow also.[49] The odor of the embryonic fluid seems to play a role in the selection of a birth site for sheep. The ewe giving birth and other ewes spend considerable time licking and sniffing the ground where the fluid has spilled at the start of the birth process.

Using deaf, anosmic (unable to smell), and blindfolded animals, I. C. Fletcher and D. R. Lindsay studied the sensory stimuli utilized by both rams and estrous ewes in partner seeking.[52] They found that the sexual activity of the rams was affected slightly by loss of hearing, more by inability to smell, and most by loss of vision. Partner seeking, on the other hand, was affected only in those rams that were unable to smell. Blindfolding and impairment of smell significantly reduced the number of ewes that mated, but only blindfolding affected the ability of the ewes to seek tethered rams.

It is noteworthy that less is known about the chemical interactions between domestic farm animals than those of many groups of free-living animals.[49] The development of interest in the behavior of domestic animals in recent years suggests that we will be in a better position in the future to determine how we should treat the animals in our care. It is generally recognized now that biostimulation is essential to the development of complete reproductive responses in domestic animals and that odor is a very important stimulant.

Applications of Animal Pheromones

Dietland Müller-Schwarze[53] says that "applications of pheromones in mammals are still at the pre-scientific stage." This is so because we lack chemical identifications of specific pheromones. Applications so far have involved the use of intra- and interspecific odors in trapping, crop production, wildlife management, and animal husbandry.

For hundreds of years trappers have used urine, scent-gland substances, and other sources of odors to attract animals.[53] Unfortunately, we lack reliable reports of controlled experiments testing the effectiveness of such practices, though many trappers make very strong claims for their methods. Evidence from experiments on other animals suggests that their claims may be true, at least in part.

According to Müller-Schwarze, gardeners and orchard caretakers often try to repel herbivores by placing the droppings of carnivores next to the crop to be protected.[53] Laboratory experiments have demonstrated that black-tailed deer feed significantly less or not at all near the feces of predators. In one experiment Müller-Schwarze tested the responses of young black-tailed deer to odors from droppings of the coyote *(Canis latrans),* the mountain lion *(Felis concolor),* the African lion *(Panthera leo),* the snow leopard *(Panthera uncia),* and the Bengal tiger *(Panthera tigris tigris).*[54] All of the droppings significantly reduced feeding by the deer. The most consistent avoidance response was elicited by the odors of the coyote and the mountain lion. Both of these predators have a distribution similar to that of the black-tailed deer. All the deer were hand-reared and had never been exposed to the actual predators. Thus Müller-Schwarze concluded, "black-tailed deer and perhaps other ungulates [hoofed animals] possess a largely genetically determined, negative response to odors of predators."

Müller-Schwarze reported that four scents are important in the social communication of the black-tailed deer.[55] The scent from the tarsal gland, located on the inner side of the hind leg, plays a primary role in mutual recognition. The scent of the forehead gland is used for marking in the environment. Urine is used to make the environment easily recognizable to the individual; it attracts or repels other individuals, depending on the circumstances. The scent produced by the metatarsal gland, located on the outside of the hind leg, acts as an alarm pheromone. Müller-Schwarze isolated an active compound from the secretion of this gland but did not succeed in identifying it. Later he suggested that identification of the alarm pheromone, which is known to inhibit feeding, might lead to its use in crop protection.[53]

In farming practice odors have been used to induce maternal behavior of domestic animals. For example, a ewe can be made to accept a strange lamb if the lamb is covered with the embryonic fluid of her own lamb or the hide of her lamb. The use of an aerosol spray containing boar odor in identifying estrous sows was mentioned above. Removal of the submaxillary gland from a young male pig makes him unable later in life to elicit the mating stance in sows and causes the females to become aggressive toward the male.[53]

This rather sketchy discussion of pheromones in animals other than insects indicates how little is known about the specific chemical compounds involved. It also indicates the widespread importance of pheromones in animal behavior. Opportunities for future research in this area are endless and exciting. Exploration is not a thing of the past, because countless frontiers remain to be explored in all kinds of research areas. The discovery of new pheromones will improve our understanding of biological phenomena and lead us to applications in agriculture that will have far-reaching, beneficial results.

CHAPTER 8

A Glimpse into the Future

INTEREST in chemical ecology is expanding rapidly. Many extremely able researchers from different disciplines are engaged in research, and soon we may expect the development of outstanding new research techniques which will enable us to attack many problems that could not be investigated previously. Moreover, our approach to investigations of chemical interactions between different types of organisms will be much more integrated.

TOWARD AN INTEGRATED CHEMICAL ECOLOGY

Already many specific chemical compounds have been identified in interactions between organisms of several different kinds. The terpenes α- and ß-pinene are volatile allelopathic agents in many species of plants.[1] They also serve as repellents to the bark beetles of several coniferous trees.[2,3] J. A. Rudinsky found that α-pinene and some other terpenes also act as attractants to the Douglas fir beetle *(Dendroctonus pseudotsugae)* in low concentrations.[4] Another terpene, 1,8-cineole, is not only a volatile allelopathic agent in several plant species[1] but also an important pollinator attractant in many species of orchids.[5]

Tannic acid is produced by many plants. Since it is toxic to most of the plants against which it has been tested, we know that it is an important allelopathic agent against higher plants. It is also very inhibitory to nitrogen-fixing and nitrifying bacteria.[1] It is, moreover, very toxic to greenbugs, and it and other tannins deter the feeding of butterflies and moths.[7] It depresses the growth of chickens,[8] rats,[9] and mice,[10] and it and other tannins often have a delayed effect on the larval growth and pupal weight of insects.[11] Most of these same effects have been demonstrated for several tannins, including some condensed tannins. All have been implicated in the resistance of some plants to viral and fungal diseases.[1]

Chlorogenic acid is commonly identified as one of the toxic compounds involved in allelopathy.[1] Todd and his colleagues[6] found that it markedly reduced the growth and survival of greenbugs. It has frequently been identified as a compound important in the resistance of plants to various diseases.[1] It also appears to be a feeding stimulant to deer in Douglas-fir foliage.[12]

p-Hydroxybenzoic acid is one of the most commonly identified benzoic-acid derivatives involved in allelopathy.[1] It also has been shown to deter the feeding of the two-spotted mite.[13]

Several long-chain fatty acids have been implicated in allelopathy among algae,[1] and several short-chain fatty acids have been demonstrated to inhibit the development and survival of the cheese mites and various insects.[14] The characteristics of rabbit odor are due to an ester of a fatty acid secreted by the anal glands.[15] The female attractants produced by male mice appear to be free fatty acids,[16] and the female attractants produced by male rats appear to be derivatives of fatty acids.[17] The vaginal discharge in rhesus monkeys contains volatile fatty acids that attract males.[18]

This cursory discussion indicates that we already have a rather primitive beginning to an integrated chemical ecology. Our knowledge of this web of chemical interactions will explode to amazing proportions in the next few decades. These early developments set the stage for exciting new applications in gardening, agriculture, forestry, horticulture, and in our general management of ecosystems.

FUTURE APPLICATIONS OF PLANT-PLANT INTERACTIONS

In a few decades we will be breeding crop plants for resistance to selected important weeds by breeding allelopathic genes into the desired cultivars, just as we now breed plants for resistance to diseases. Wild type relatives of crop plants will be screened for their allelopathic potential against selected weeds. If promising types will not hybridize with desirable cultivars, genetic engineering techniques, such as protoplast fusion, will be used to introduce the allelopathic genes into the cultivars. Wild types will also be selected for resistance to the allelopathic agents produced by certain important weed species so that these resistances may be introduced into desirable cultivars.

Of course, screening of various crop varieties and cultivars will be expanded to select those with allelopathic potential or resistance to allelopathic agents, or both. Widespread use of mulches of allelopathic crop plants and weeds will help control weeds in crops. Interference—that is, allelopathy plus competition—by some weeds can reduce crops yields 100 percent in some conditions. Reductions in yields of 40 to 80 percent are common.[19, 20] In many cases, elimination of weed interference for only three to four weeks after emergence of the crop plants is almost as effective as weed control throughout the growing season.[21, 22] An initial mulch application can suffice. Weed control will be revolutionized by such knowledge and by research into the relative impacts of the allelopathic and competitive components of weed interference.

New allelopathic agents will be identified from plants with allelopathic potential, and some of these compounds will be used as phytocides in place of many of the synthetic weed killers used at present. Toxins produced by fungal pathogens will probably be used widely for control of some weeds. Already the toxins produced by fungal pathogens of Canada thistle are being tested for control of this weed, and three toxins have proved promising.[23] These toxins can work even if the weeds are resistant to infection by the fungi that produce them. Use of natural chemical interactions along with other biological control mechanisms, such as weed predators and diseases, will virtually eliminate our need for synthetic weed killers.

Some weeds produce compounds that stimulate the growth of crop plants.[24] The application to wheat fields of agrostemmin, a compound produced by corn cockle, at the extremely low rate of 1.2 grams per hectare increased wheat yields on both fertilized and unfertilized areas. Many other compounds identified from weeds will increase crop yields in the future.

It is common now in many developing countries to plant several crop plants in the same fields in order to increase total crop yields. Apparently chemicals produced by some crop plants stimulate others. S. F. Zabyalyendzik[25] found this to be true under greenhouse and field conditions for some crop plants. He investigated the mutual interactions of buckwheat, lupine *(Lupinus),* mustard, and oats and reported that yields of buckwheat tops were 30 to 35 percent greater, and grain yields 12 to 35 percent greater, in mixtures with other components than in pure stands. He reported that water-soluble root exudates of lupine and mustard stimulated growth and development of buckwheat.

As commercial fertilizers become more expensive, there will be increased use of legumes for biological nitrogen fixation, and many legumes are known to be allelopathic to several crop plants, as shown in Chapter 3. For example, D. F. Lykhvar and N. S. Nazarova investigated the growth of several species of legumes and corn in pure and mixed cultures. They reported that the beneficial effects of growing legumes in mixed cultures with corn depended on the specific varieties of the legume species.[26] Many varieties had a detrimental effect in mixed cultures, indicating negative allelopathic effects. Subsequently, new varieties of legumes were developed specifically for use in mixed cultures with corn or other crop plants. In the future there will be great emphasis on selection and breeding of compatible plants for mixed cropping throughout the world.

FUTURE APPLICATIONS OF PLANT-NEMATODE INTERACTIONS

The success attained in controlling plant-parasitic nematodes by interplanting or rotation of certain resistant species with susceptible crop plants will lead to much greater use of these techniques by gardeners and farmers. The effectiveness of certain species of marigold, chrysanthemum, rattlebox, margosa, and castor bean has been demonstrated against several nematodes. Other less widely tested plants, such as asparagus, effectively lower parasitic nematode populations in some cases. Many other crop plants will be found to have this capacity. Others will be bred for resistance to nematodes by introducing genes that cause the production of repellents, deterrents, or chemicals with delayed effects in the crop plants.

FUTURE APPLICATIONS OF PLANT-INSECT INTERACTIONS

Plants in natural ecosystems are dependent entirely on their own defenses against insects and other herbivores. Crop plants have been selected and bred for elimination or reduced concentrations of many of the defensive chemical compounds because they taste bitter. Nonetheless, most crop plants have at least some resistance to some insect pests, and varieties and related species can often be found that have considerable resistance to selected pests (see Chapters 6 and 7). These can be used in breeding programs to make varieties with other desirable characteristics resistant, or at least less susceptible, to the same pests. This technique has been used very little for control of insects and other herbivore pests in the past, probably because of the rapid development of many effective commercial pesticides. In the future there will be a tremendous increase in the breeding of crop plants resistant to animal pests. We might even match the resistance that we have achieved in breeding plants resistant to plant diseases. Approximately 75 percent of the current agricultural acreage in the United States is planted with varieties resistant to one or more plant diseases.

A high percentage of crop plants is grown for production of fruits and seeds. These plants could be bred for high concentrations of chemical repellents or deterrents in vegetative structures and for low concentrations in the fruits and seeds, if the compounds involved are undesirable in the fruits and seeds.

It is likely that systemic repellents will be found or synthesized that will remain effective during most of the growing season and will continue to move into the new growth of the crop plants. For example, a repellent will be found that will keep commercial honeybees away from vegetation recently sprayed with a dangerous pesticide. Possibly some of the repellents among the defensive insect secretions will prove to be effective when sprayed on plants.

Often there is a close synchrony between the developmental stages of insect pests and their host plants. D. C. Eidt and C. H. A. Little[27] suggested in 1968 that sprays of plant hormones (such as kinins, gibberellins, and auxins) might disrupt the synchrony. A short time later P. Scholze[28] proved that aphids that had fed on kinetin-treated plant parts of broad bean *(Vicia faba)* weighed less than those that fed on untreated parts. Subsequently, S. Scheurer[29] found that female aphids *(Aphis fabae)* that had fed on the leaves or roots of broad-bean plants previously treated with kinetin produced few larvae or did not reproduce at all. This application of plant hormones is not economically feasible at present, but future research could certainly make it so, particularly for special cases where other control mechanisms have not proved effective.

FUTURE APPLICATIONS OF INSECT-INSECT INTERACTIONS

Sex-attractant pheromones are being identified at a rapid pace, pointing the way toward expanded use of these compounds in pest control. Various uses were described in Chapter 6, and others will be devised. The small amounts required and the general absence of detrimental side effects make these compounds highly useful in pest control. Since these chemicals are most effective on low populations of pests, they will make it possible to eliminate some very serious pests completely in selected regions. Sex pheromones will become more and more important in controlling pests that have become resistant to conventional pesticides.

Much more effective slow-release agents, humectants, and diluents will be discovered for use with sex attractants, and much more effective and economical methods of dispersing the formulations will be devised. In those cases where the sex pheromones are used as lures rather than to disrupt insect guidance systems, more effective methods will be developed to eliminate the attracted pests. Pheromones will also be used to attract insect pests to sources of pathogenic bacteria, fungi, viruses, or protozoa. The attracted pests will carry the diseases back to their colonies and spread it among their populations, thus eliminating many of their species.

A few years ago the future looked very bright for the use of juvenile hormones and ecdysones in insect pest control. The idea was exciting because juvenile hormones are restricted to insects and have no known effects on other animals or plants.[30] Moreover, many compounds with juvenile-hormone activity have been isolated from plants, particularly from conifers. Many ecdysones have been identified, as described in Chapter 6—more from plants than from animals.[31] These molting hormones are generally not as active as the juvenile hormones, and they are expensive to synthesize because they have a steroid nucleus. They also penetrate the insect cuticle less readily. Many insects synthesize substances that regulate their cuticular permeability, and some of those compounds will, no doubt, be identified and used along with the ecdysones to promote penetration of the ecdysones.[32] The ecdysones can disrupt development at any larval or pupal stage. They also act as female chemosterilants in insects and reduce egg hatch.

At present the uses of juvenile hormones and ecdysones in pest control are difficult to predict. The Zoecon Corporation has been granted an experimental permit by the United States Environmental Protection Agency to test a juvenile hormone analogue (Altosid®) as a larvacide for floodwater mosquitoes. This appears to be an encouraging development.[32] In general, it now seems that synthetic juvenile hormones will be used in agriculture in a limited way, but synthetic ecdysones hold little promise for use even on a relatively small scale, though they may be used in special cases as female chemosterilants and to reduce egg hatch. Both kinds of hormones will be involved in the breeding of at least some crop species for greater insect resistance. It is to be hoped that future research and development will prove these predictions wrong.

The chemicals produced by animals that influence the behavior of other animal species are called *allomones*. In the future these will probably play important roles in the control of insect pests. Both parasitic and predatory insects are now used to control certain insect pests, and the chemicals produced by the host insects are known to play a major role in host selection by the parasites and predators. Thus these chemicals can be used to manage parasites so that they are more effective in pest control.[33] Four phases have been defined in the host-selection process: finding the host habitat, finding the host, acceptance of the host by the parasite, and determination of host suitability. Often olfactory cues produced by the host are involved in all four phases, along with visual and physical cues in some or all phases.

Considerable progress has been made in identifying the chemical compounds involved in host selection. There will be a tremendous acceleration in the identification of these compounds and in the production of synthetic compounds. Thus we will see a rapid increase in the use of parasitic and predatory insects in the control of insect pests.

FUTURE USES OF OTHER CHEMICAL INTERACTIONS IN AGRICULTURE

Plant-Animal Interactions. A few examples of the damage done by small animals to cultivated crops, orchards, and forests may be of interest. One ground squirrel is capable of destroying over 50 pounds of wheat per season. A population of twelve squirrels per acre may decrease the annual dry forage yield by more than 1,000 pounds per acre, or about 38 percent. One hundred meadow mice per acre can consume 4 percent of the annual production in alfalfa. Marmots or woodchucks may eat a pound of green food daily and destroy more by trampling in their runways.[34] Pocket gophers have been found to destroy from one-fourth to one-third of the alfalfa crop over a large area.

As shown in Chapter 7, our knowledge of chemical interactions between crop plants and animals other than nematodes and insects is fragmentary. Because interest in this area is growing, the pace of research should accelerate in coming decades. Interesting new applications of our knowledge are regularly being made around the world. Recently M. Takahashi reported the effects of cluster amaryllis *(Lycoris radiata)* planted on the small levees of rice fields in Japan to keep mice out of the rice paddies.[35] Investigations demonstrated that mice dug through a 10-centimeter dirt wall in 27 minutes on the average, but failed to penetrate the same type of wall when cluster amaryllis was planted in it. Mice avoided soil on which juice from amaryllis bulbs was sprinkled, and experiments indicated that an alkaloid in the bulbs keeps mice away.

Cluster amaryllis have beautiful flowers and thus add beauty to the fields in addition to keeping out mice. A great many border plants will be found that can be used in this manner around fields and gardens to keep undesired animals away. Up to this time we have emphasized the use of synthetic chemicals to kill undesired herbivore pests rather than trying to keep them out of selected areas. As costs and our knowledge of the undesirable side effects of synthetic chemicals increase, more and more emphasis will be placed on natural methods of control.

There will definitely be increasingly selective breeding of crop plants and forest trees that are resistant to the animal pests in a given region, such as deer, rabbits, hares, snails, slugs, voles, and mites. Fortunately, we can select for many types of chemical compounds in plants in order to prevent grazing. Among these are alkaloids, certain lactones, fatty acids, cyanogenic glycosides, cyanogenic lipids, tannins, certain free amino acids, essential oils, saponins, and several phenolic compounds.

Animal-Animal Interactions. Chapter 7 shows our knowledge of animal communication by pheromones is not very extensive except among the insects. This is unfortunate because many very important applications could be made in agriculture if more information were available.

We will soon identify many of the space-marking pheromones of the serious small-mammal pests of garden and crop plants, such as rabbits, mice, and rats. Mixtures of the pheromones, applied to gardens or fields, will disrupt the behavior of the animals and keep them out of the marked areas. In many cases it will be necessary to spray the pheromones only around the margins of the gardens or fields.

As described in Chapter 7, mice produce pheromones that induce fear in other mice. This is probably true of at least a few other animal pests. Identification of the pheromones will give us an excellent tool for excluding certain pests from garden and field crops. Many of these compounds will be identified soon.

The majority of small mammals live either beneath the surface of the ground or on the surface where denseness of vegetation or other barriers prevent visual or acoustic communication except when the animals are at close range. Thus sex-attractant pheromones are very important to the reproduction of these animals. Many of these attractants will be identified in the next decade and will be used to control reproduction in these mammal pests in ways similar to those used in controlling insects. Some will be used to confuse the mammals and keep them from finding mates. Others will be used to trap unwanted mammals or to attract them to poison bait stations. In some countries losses of garden and field crops caused by small mammals are very great. The use of pheromones to control the pest populations in these cases will be widespread and extremely important economically.

Huge losses occur in the storage of grains and other food materials in some countries of the world because of various pests, including small mammals. Undoubtedly several of the pheromones will be used widely to prevent this destruction of critical food supplies.

Chapter 7 also describes the use of the droppings of large predators, such as coyotes and mountain lions, to keep deer out of gardens and fields. Within the next decade several important territorial marking pheromones, fear pheromones, and other types of pheromones will be identified and synthesized for use in repelling various agricultural pests.

Unfortunately, much less is known about the roles and identifications of pheromones in domestic animals than in wild animals. Some progress has been made. As reported in Chapter 7, aerosols containing steroid compounds from the saliva of boars are used now to detect estrous sows. Even though, under domestication, there has been a selection against many forms of behavior that depend on communication by odors, it is still generally realized that biostimulation is essential to the development of complete reproductive responses, and that odor is a very important stimulant. There will be a pronounced acceleration in the rate of research on pheromones produced by domestic animals, and this will result in a much better understanding of how we should treat animals in our care. There are often rather serious behavior patterns in certain domestic animals that, for example, cause mothers to kill their young. Undoubtedly, pheromones will be identified that will prevent such an occurrence. Some of the most exciting new discoveries in the field of chemical ecology will occur in this area.

In overall summary, the coming decades will be exciting in all areas of chemical ecology. There is a growing awareness in many scientific disciplines of the enticing research opportunities and important applications connected with allelochemics and pheromones. Some of the scientific disciplines involved include ecology, physiology, agronomy, pomology, viticulture, horticulture, weed science, biochemistry, animal husbandry, diary science, forage and range science, analytical chemistry, organic chemistry, phycology, pathology, mycology, zoology, forestry, and entomology. Obviously, there are overlaps in this list and omissions. The interest and dedication of scientists in many fields will make possible the rapid advances that lie ahead.

NOTES

CHAPTER 1

1. Theophrastus, *Enquiry into Plants and Minor Works on Odours and Weather Signs,* A. Hort (London: W. Heinemann, 1916), vol. 2, book 8.

2. Plinius Secundus, C., *Natural History,* trans. H. Rackam, W. H. S. Jones, and D. E. Eichholz (Cambridge, Mass.: Harvard University Press, 1938-63).

3. Dioscorides, P., *Greek Herbal,* trans. J. Goodyer, ed. R. T. Gunther (New York: Hafner Pub. Co., 1959).

4. Gerarde, J., *The Herball or Generall Historie of Plantes* (London: John Norton, 1597).

5. Culpeper, Nicholas, *English Physitian and Complete Herball* (London: W. Foulsham, 1633).

6. *The Works of Sir Thomas Browne,* ed. Keynes, G. (London: Faber & Gwyer, 1929), vol. 4.

7. Lee, I. K., and M. Monsi, "Ecological studies on *Pinus densiflora* forest 1. Effects of plant substances on the floristic composition of the undergrowth," *Bot. Mag.* (Tokyo) 76 (1963): 400-13.

8. Candolle, Augustin Pyrame de, *Physiologie végétale,* (Paris: Béchet Jeune, Lib. Fac. Méd., 1832), 3: 1474-75.

9. "Editor's Note," *Gard. Chron.* 1 (1841): 801.

10. T. T., "Beech-trees," *Gard. Chron.* 2 (1842): 253.

11. Montrose Inquirer, "Fairy rings," *Gard. Chron.* 5 (1845): 722.

12. Way, J. T., "On the fairy-rings of pastures as illustrating the use of inorganic manures." *J. Roy. Agr. Soc.* 7 (1847): 549-52.

13. Beobachter, "The highland pine," *Gard. Chron.* 5 (1845): 69-70.

14. T. A., "Rotation of crops," *Gard. Chron.* 5 (1845): 159.
15. Editorial, "Clover sickness," *Gard. Chron.* 7 (1847): 41.
16. Young, A., *The Farmers Calendar* (London: 1804).
17. Stickney, J. S., and P. R. Hoy, "Toxic action of black walnut," *Trans. Wisc. State Hort. Soc.* 11 (1881): 166-67.
18. Willis, J. J., letter, *Gard. Chron.*, 3d ser., 16 (1894): 98-99.
19. Willis, J. J., "Peppermint culture," *Gard. Chron.*, 3d ser., 16 (1894): 594.
20. Webster, Mr., "Question Box: Can crops of any kind be grown near to small fruits without injury to them?" *Trans. Ill. Hort. Soc.*, n.s., 27 (1893): 243.
21. Austin, Mr., "Discussion of questions from the question box," *Trans. Ill. Hort. Soc.*, n.s., 29 (1895): 233.
22. Rice, E. L., *Allelopathy* (New York: Academic Press, 1974).
23. Rice, E. L., "Allelopathy—an update," *Bot. Rev.* 45 (1979): 15-109.
24. Schreiner, O., and H. S. Reed, "The production of deleterious excretions by roots," *Bull. Torrey Bot. Club* 34 (1907): 279-303.
25. Schreiner, O., and H. S. Reed, "The toxic action of certain organic plant constituents," *Bot. Gaz.* 45 (1908): 73-102.
26. Schreiner, O. and M. X. Sullivan, "Soil fatigue caused by organic compounds," *J. Biol. Chem.* 6 (1909): 39-50.
27. Pickering, S. V., "The effect of one plant on another," *Ann. Bot.* 31 (1917): 181-87.
28. Pickering, S.V., "The action of one crop on another," *J. Roy. Hort. Soc.* 43 (1919): 372-80.
29. Massey, A. B., "Antagonism of the walnuts (*Juglans nigra* L. and *J. cinerea* L.) in certain plant associations," *Phytopathology* 15 (1925): 773-84.
30. Davis, R. F., "The toxic principle of *Juglans nigra* as identified with synthetic juglone and its toxic effects on tomato and alfalfa plants," *Amer. J. Bot.* 15 (1928): 620.
31. Elmer, O. H., "Growth inhibition of potato sprouts by the volatile products of apples," *Science* 75 (1932): 193.
32. Molisch, H., *Der Einfluss einer Pflanze auf die andere-Allelopathie* (Jena: Gustav Fischer Verlag, 1937).
33. Bode, H. R., "Über die Blattausscheidungen des Wermuts und ihre Wirkung auf andere Pflanzen," *Planta* 30 (1940): 567-89.
34. Funke, G. L., "The influence of *Artemisia absinthium* on neighbouring plants," *Blumea* 5 (1943): 281-93.
35. Benedict, H. M., "The inhibitory effect of dead roots on the growth of bromegrass," *J. Amer. Soc. Agron.* 33 (1941): 1108-1109.

CHAPTER 2

1. Theophrastus, *Enquiry into Plants and Minor Works on Odours*

and Weather Signs, trans. A. Hort (London: W. Heinemann, 1916).
2. Plinius Secundus, C., *Natural History,* trans. H. Rackam, W. H. S. Jones, and D. E. Eichholz (Cambridge, Mass.: Harvard University Press, 1938-63).
3. Gerarde, J., *The Herball or Generall Historie of Plantes* (London: John Norton, 1597).
4. Culpeper, Nicholas, *English Physitian and Complete Herball* (London: W. Foulsham, 1633).
5. Coles, W., *Adam in Eden: or Natures Paradise* (London: N. Brooke, 1657).
6. Culpeper, N., *The English Physician Enlarged* (London: J. Barker, 1681).
7. Beinhart, E. G., "Production and use of nicotine," in United States Department of Agriculture, *The Yearbook of Agriculture: Crops in Peace and War* (Washington, D.C.: Government Printing Office, 1950-51), pp. 772-79.
8. Eliot, J., *Essays upon Field-Husbandry in New-England, as It Is or May Be Ordered* (Boston: Edes and Gill, 1760).
9. Deane, S., *The New England Farmer; or Georgical Dictionary* (Worcester, Mass.: Isaiah Thomas, 1790).
10. Editorial comments, "Elder leaves," *New-York Farmer and Amer. Gard. Mag.* 3 (1830): 143.
11. Rafinesque, C. S., *Medical Flora; or Manual of Medical Botany of the United States of North America,* (Philadelphia: Atkinson, 1830), vol. 2.
12. Miscellaneous, "Acorn squash," *New-York Farmer and Amer. Gard. Mag.* 3 (1830): 26.
13. Editorial Comments, "Garlic," *New-York Farmer and Amer. Gard. Mag.* 7 (1834): 247.
14. Mackenzie, P., "The bear's-foot *(Helleborus foetidus),"* *Gard. Chron.* 5 (1845): 594.
15. Editor's Note, "*Ailanthus* bark," *Gard. Chron.,* 3d ser., 18 (1895): 619.
16. Crosby, D. G., "Minor insecticides of plant origin," *Naturally Occurring Insecticides,* ed. M. Jacobson and D. G. Crosby (New York: Marcel Dekker, 1971), pp. 177-239.
17. Aristotle, *Historia Animalium,* trans. D'Arcy Thompson (London: Oxford University Press, 1910), p. 624.
18. Dobbs, A., "Concerning bees, and their method of gathering wax and honey," *Philos. Trans. Roy. Soc.* 46 (1750):536-49.
19. Huber, F., *New Observations on the Natural History of Bees,* 3d ed. (London: W. and C. Tait, and Longman, Hurst, Rees, Orne, and Brown, 1821).
20. Darwin, C. R., *On the Various Contrivances by Which British and Foreign Orchids Are Fertilised by Insects and on the Good Effects of Intercrossing* (London: J. Murray, 1862).

21. Darwin, C. R., *The Effects of Cross and Self Fertilization in the Vegetable Kingdom* (New York: D. Appleton, 1902).
22. Rodway, J., *In the Guiana Forest; Studies of Nature in Relation to the Struggle for Life,* 2d ed. (London: T. Fisher Unwin, 1895).
23. Bethe, A., "Vernachlässigte Hormone," *Naturwissenschaften* 20 (1932): 177-81.
24. Karlson, P., and A. Butenandt, "Pheromones (Ectohormones) in Insects," *Annu. Rev. Ent.* 4 (1959): 39-58.
25. Whittaker, R. H., "The biochemical ecology of higher plants," in *Chemical Ecology,* ed. E. Sondheimer and J. B. Simeone, pp. 43-70. (New York: Academic Press, 1970).

CHAPTER 3

1. McCalla, T. M. and F. L. Duley, "Stubble mulch studies: Effect of sweetclover extract on corn germination," *Science* 108 (1948): 163.
2. McCalla, T. M., and F. L. Duley, "Stubble mulch studies: III. Influence of soil microorganisms and crop residues on the germination, growth and direction of root growth of corn seedlings." *Soil Sci. Soc. Amer. Proc.* 14 (1949): 196-99.
3. Norstadt, F. A., and T. M. McCalla, "Phytotoxic substance from a species of *Penicillium,*" *Science* 140 (1963): 410-11.
4. Ellis, J. R., and T. M. McCalla, "Effects of patulin and method of application on growth stages of wheat," *Appl. Microbiol.* 25 (1973): 562-66.
5. Guenzi, W. D., and T. M. McCalla, "Phytotoxic substances extracted from soil," *Soil Sci. Soc. Amer. Proc.* 30 (1966): 214-16.
6. Patrick, Z. A., and L. W. Koch, "Inhibition of respiration, germination and growth by substances arising during the decomposition of certain plant residues in the soil," *Canad. J. Bot.* 36 (1958): 621-47.
7. Patrick, Z. A., T. A. Toussoun, and W. C. Snyder, "Phytotoxic substances in arable soils associated with decomposition of plant residues," *Phytopathology* 53 (1963): 152-61.
8. Burgos-Leon, W. "Phytotoxicité induite par les résidus de récolte de *Sorghum vulgare* dans les sols sableux de l'ouest Africain," Thèse pour Doctorat, Université de Nancy, France, 1976.
9. Stevenson, F. J., "Organic acids in soil," in *Soil Biochemistry,* ed. A. D. McLaren and G. H. Peterson (New York: Marcel Dekker, 1967), pp. 119-42.
10. Chou, C. H., and H. J. Lin, "Autointoxication mechanisms of *Oryza sativa.* I. Phytotoxic effects of decomposing rice residues in soil," *J. Chem. Ecol.* 2 (1976): 353-67.
11. Huang, C. Y., "Effects of nitrogen fixing activity of blue-green algae on the yield of rice plants," *Bot. Bull. Academia Sinica* 19 (1978): 41-52.

NOTES

12. Galston, A. W., "The water fern-rice connection," *Natural Hist.* 12 (1975): 10.
13. Chou, C. H., T. J. Lin, and C. I. Kao, "Phytotoxins produced during decomposition of rice stubbles in paddy soil and their effect on leachable nitrogen," *Bot. Bull. Academia Sinica* 18 (1977): 45-60.
14. Rice, E. L., C. Y. Lin, and C. Y. Huang, "Effects of decomposing rice straw on growth and nitrogen fixation by *Rhizobium*," *J. Chem. Ecol.* 7 (1981): 333-44.
15. Rice, E. L., C. Y. Lin, and C. Y. Huang, "Effects of decaying rice straw on growth and nitrogen fixation of a bluegreen alga," *Bot. Bull. Academia Sinica* 21 (1980): 111-17.
16. Menzies, J. D., and R. G. Gilbert, "Responses of the soil microflora to volatile components in plant residues," *Soil Sci. Soc. Amer. Proc.* 31 (1967): 495-96.
17. Dadykin, V. P., L. N. Stepanov, and B. E. Ryzhkova, "On the importance of volatile plant secretions under the development of closed systems," in *Physiological-Biochemical Basis of Plant Interactions in Phytocenoses,* ed. A. M. Grodzinsky (Kiev: Naukova Dumka, 1970), 1: 118-24 (in Russian).
18. Petrova, A. G., "Effect of phytoncides from soybean, gram chickpea and bean on the uptake of phosphorus by maize," in *Interactions of Plants and Microorganisms in Phytocenoses,* ed. A. M. Grodzinsky (Kiev: Naukova Dumka, 1977), pp. 91-97 (in Russian, English summary).
19. Kozel, P. C., and H. B. Tukey, Jr., "Loss of gibberellins by leaching from stems and foliage of *Chrysanthemum morifolium* 'Princess Anne,'" *Amer. J. Bot.* 55 (1968): 1184-89.
20. Tamura, S., C. Chang, A. Suzuki, and S. Kumai, "Isolation and structure of a novel isoflavone derivative in red clover," *Agric. Biol. Chem.* 31 (1967): 1108-1109.
21. Tamura, S., C. Chang, A. Suzuki, and S. Kumai, "Chemical studies on 'clover sickness'. Part I. Isolation and structural elucidation of two new isoflavonoids in red clover," *Agric. Biol. Chem.* 33 (1969): 391-97.
22. Chang, C., A. Suzuki, S. Kumai, and S. Tamura, "Chemical studies on 'clover sickness'. II. Biological functions of isoflavonoids and their related compounds," *Agric. Biol. Chem.* 33 (1969): 398-408.
23. Katznelson, J., "Studies in clover soil sickness. I. The phenomenon of soil sickness in berseem and Persian clover," *Pl. & Soil* 36 (1972): 379-93.
24. Rakhteenko, I. N., I. A. Kaurov, and I. T. Minko, "Effect of water-soluble metabolites of a series of crops on some physiological processes," in *Physiological-Biochemical Basis of Plant Interactions in Phytocenoses,* ed. A. M. Grodzinsky (Kiev: Naukova Dumka, 1973), 4: 23-26 (in Russian, English summary).
25. Grümmer, G., and H. Beyer, "The influence exerted by species of *Cammelina* on flax by means of toxic substances," in *The Biology of Weeds,* Symposium British Ecol. Soc., ed. J. L. Harper, (Oxford: Black-

well Scientific Pub., 1960), pp. 153-57.

26. Bieber, G. L., and C. S. Hoveland, "Phytotoxicity of plant materials on seed germination of crownvetch, *Coronilla varia* L.," *Agron. J.* 60 (1968): 185-88.

27. Schreiber, M. M., and J. L. Williams, Jr., "Toxicity of root residues of weed grass species," *Weeds* 15 (1967): 80-81.

28. Bell, D. T., and D. E. Koeppe, "Noncompetitive effects of giant foxtail on the growth of corn," *Agron. J.* 64 (1972): 321-25.

29. Gressel, J. B., and L. G. Holm, "Chemical inhibition of crop germination by weed seeds and the nature of inhibition by *Abutilon theophrasti,*" *Weed Res.* 4 (1964): 44-53.

30. Anaya, A. L., and A. Gomez-Pompa, "Inhibicion del crecimiento producida por el 'piru' (*Schinus molle* L.)," *Revista Soc. Mex. Hist. Nat.* 32 (1971): 99-109.

31. Salas, M. C., and E. Vieitez, "Actividad de crecimiento de Ericaceas." *Anales Edafol. Agrobiol.* 31 (1972): 1001-1009.

32. Ballester, A., and E. Vieitez, "Estudio de sustancias de crecimiento aisladas de *Erica cinerea* L." *Acta Ci. Compostelana* 8 (1971): 79-84.

33. Einhellig, F. A., and J. A. Rasmussen, "Allelopathic effects of *Rumex crispus* on *Amaranthus retroflexus,* grain sorghum and field corn," *Amer. Midl. Naturalist* 90 (1973): 79-86.

34. Tames, R. S., M. D. V. Gesto, and E. Vieitez, "Growth substances isolated from tubers of *Cyperus esculentus* var. *aureus,*" *Physiol. Pl.* 28 (1973): 195-200.

35. Buchholtz, K. P., "The influence of allelopathy on mineral nutrition," in Environmental Physiol. Subcomm., U.S. Nat. Comm. for IBP, *Biochemical Interactions Among Plants* (Washington, D.C.: National Academy of Science, 1971), pp. 86-89.

36. Minar, J., "The effect of couch grass on the growth and mineral uptake of wheat," *Folia Fac. Sci. Nat. Univ. Purkynianae Brun.* 15 (1974): 1-84.

37. Bendall, G. M., "The allelopathic activity of California thistle (*Cirsium arvense* (L.) Scop.) in Tasmania," *Weed Res.* 15 (1975): 77-81.

38. Rasmussen, J. A., and F. A. Einhellig, "Noncompetitive effects of common milkweed, *Asclepias syriaca* L., on germination and growth of sorghum," *Amer. Midl. Naturalist* 94 (1975): 478-83.

39. Wilson, R. E., and E. L. Rice, "Allelopathy as expressed by *Helianthus annuus* and its role in old-field succession," *Bull. Torrey Bot. Club* 95 (1968): 432-48.

40. Gajić, D., "Interaction between wheat and corn cockle on brown soil and smonitsa," *J. Sci. Agric. Res.* 19 (1966): 63-96.

41. Gajić, D., and M. Vrbaski, "Identification of the effect of bioregulators from *Agrostemma githago* upon wheat in heterotrophic feeding, with special respect to agrostemmin and allantoin," *Fragm. Herb. Croatica* 7 (1972): 1-6.

42. Gajić, D., S. Malencić, M. Vrbaski, and S. Vrbaski, "Study of the

possible quantitative and qualitative improvement of wheat yield through agrostemin as an allelopathic factor," *Fragm. Herb. Jugoslavica* 63 (1976): 121-41.

43. Peters, E. J., "Toxicity of tall fescue to rape and birdsfoot trefoil seeds and seedlings," *Crop Sci.* 8 (1968): 650-53.

44. Dzubenko, N. N., and N. I. Petrenko, "On biochemical interaction of cultivated plants and weeds," in *Physiological-Biochemical Basis of Plant Interactions in Phytocenoses,* ed. A. M. Grodzinsky (Kiev: Naukova Dumka, 1971), 2: 60-66 (in Russian, English summary).

45. Neustruyeva, S. N., and T. N. Dobretsova, "Influence of some summer crops on white goosefoot," in *Physiological-Biochemical Basis of Plant Interactions in Phytocenoses,* ed. A. M. Grodzinsky (Kiev: Naukova Dumka, 1972), 3: 68-73 (in Russian, English summary).

46. Markova, S. A., "Experimental investigations of the influence of oats on growth and development of *Erysimum cheiranthoides* L.," in *Physiological-Biochemical Basis of Plant Interactions in Phytocenoses,* ed. A. M. Grodzinsky (Kiev: Naukova Dumka, 1972), 3: 66-68 (in Russian, English summary.)

47. Prutenskaya, N. I., "Peculiarities of interaction between *Sinapis arvensis* L. and cultivated plants," in *Physiological-Biochemical Basis of Plant Interactions in Phytocenoses,* ed. A. M. Grodzinsky (Kiev: Naukova Dumka, 1974), 5: 66-68 (in Russian, English summary).

48. Gajić, D., "The effect of agrostemins as a means of improvement of the quality and quantity of the grass-cover of the Zlatibor—as a preventive measure against the weeds," paper delivered at Yugoslav Symposium on Weed Control in Hilly and Mountainous Areas, Sarajevo, 1973.

49. Putnam, A. R., and W. B. Duke, "Biological suppression of weeds: evidence for allelopathy in accessions of cucumber," *Science* 185 (1974): 370-72.

50. Fay, P. K., and W. B. Duke, "An assessment of allelopathic potential in *Avena* germplasm," *Weed Sci.* 25 (1977): 224-28.

51. Panchuk, M. A., and N. I. Prutenskaya, "On the problem of the presence of allelopathic properties in wheat-wheat grass hybrids and their initial forms," in *Physiological-Biochemical Basis of Plant Interactions in Phytocenoses,* ed. A. M. Grodzinsky (Kiev: Naukova Dunka, 1973), 4: 44-47 (in Russian, English summary).

52. Grodzinsky, A. M., and M. A. Panchuk, "Allelopathic properties of crop residues of wheat-wheat grass hybrids," in *Physiological-Biochemical Basis of Plant Interactions in Phytocenoses,* ed. A. M. Grodzinsky (Kiev: Naukova Dumka, 1974), 5: 51-55 (in Russian, English summary).

53. Atsatt, P. R., "Biochemical bridges between vascular plants," *Biochemical Coevolution,* ed. K. L. Chambers, Biol. Colloquium no. 29 (Corvallis: Oregon State University Press, 1970), pp. 53-68.

54. Graham, B. F., Jr., and F. H. Bormann, "Natural root grafts," *Bot. Rev.* 32 (1966): 255-92.

55. Verrall, A. F., and T. W. Graham, "The transmission of *Ceratosto-*

mella ulmi through root grafts," *Phytopathology* 25 (1935): 1039-40.

56. Björkman, E., "*Monotropa hypopitys* L.—an epiparasite on tree roots," *Physiol. Plant.* 13 (1960): 308-27.

57. Wilde, S. A., and A. Lafond, "Symbiotrophy of Lignophytes and fungi: its terminological and conceptual deficiencies," *Bot. Rev.* 33 (1967): 99-104.

58. Woods, F. W., and K. Brock, "Interspecific transfer of Ca^{45} and P^{32} by root systems," *Ecology* 45 (1964): 886-89.

59. Bell, A. A., "Plant pathology as influenced by allelopathy," in *Report of the Research Planning Conference on the Role of Secondary Compounds in Plant Interactions (Allelopathy),* ed C. G. McWhorter, A. C. Thompson, and E. W. Hauser (Tifton, Ga.: Agricultural Research Service, USDA, 1977), pp. 64-99.

60. Stotzky, G., and S. Schenck, "Observations on organic volatiles from germinating seeds and seedlings," *Amer. J. Bot.* 63 (1976): 798-805.

61. Allen, R. N., and F. J. Newhook, "Suppression by ethanol of spontaneous turning activity in zoospores of *Phytophthora cinnamomi,*" *Trans. Brit. Mycol. Soc.* 63 (1974): 383-85.

62. Mishra, R. R., and K. K. Pandey, "Studies on soil fungistasis: V. Effect of temperature, moisture content and incubation period," *Indian Phytopathol.* 27 (1974): 475-79.

63. Mishra, R. R. and K. K. Pandey, "Studies on soil fungistasis IV. Effect of physico-chemical characters and soil fungal flora on fungistasis," *Ann. Edafol. Agrobiol.* 34 (1975): 423-28.

64. Li, C. Y., K. C. Lu, E. E. Nelson, W. B. Bollen and J. M. Trappe, "Effect of phenolic and other compounds on growth of *Poria weirii* in vitro," *Microbios* 1 (1969): 305-311.

65. Li, C. Y., K. C. Lu, J. M. Trappe and W. B. Bollen, "Separation of phenolic compounds in alkali hydrolysates of a forest soil by thin-layer chromatography," *Canad. J. Soil Sci.* 50 (1970): 458-60.

66. Li, C. Y., K. C. Lu, J. M. Trappe and W. B. Bollen, "*Poria weirii*-inhibiting and other phenolic compounds in roots of red alder and Douglas-fir," *Microbios* 5 (1972): 65-68.

67. Li, C. Y., et al., "Formation of p-hydroxybenzoic acid from phenylacetic acid by *Poria weirii,*" *Canad. J. Bot.* 51 (1973): 827-28.

68. Baker, K. F., and W. C. Snyder, eds., *Ecology of Soil-Borne Plant Pathogens* (Berkeley: University of California Press, 1965).

69. Baker, K. F., and R. J. Cook, *Biological Control of Plant Pathogens* (San Francisco: W. H. Freeman, 1974).

70. Bruehl, G. W., ed., *Biology and Control of Soil-Borne Plant Pathogens.* (St. Paul, Minn.: American Phytopathological Society, 1975).

71. McGrath, W. T., "Biological control of *Fomes annosus:* a new possibility in the United States," *Consultant* 17 (1972): 94-96.

72. Yakhontov, A. F., "On the possibility of using the allelopathic action of some plants for controlling *Dactilospheara viticola* F.," in *Physio-*

logical-Biochemical Basis of Plant Interactions in Phytocenoses, ed. A. M. Grodzinsky (Kiev: Naukova Dumka, 1973), 4: 57-60.

73. Wood, R. K. S., A. Ballio, and A. Graniti, eds., *Phytotoxins in Plant Disease* (New York: Academic Press, 1972).

74. Strobel, G. A., "Phytotoxins produced by plant parasites," *Annual Rev. Plant Physiol.* 25 (1974): 541-66.

75. Farkas, G. L. and Z. Kiraly, "Role of phenolic compounds in the physiology of plant diseases and disease resistance," *Phytopathol. Z.* 44 (1962): 105-150.

76. Bell, A. A., "Biochemical bases of resistance of plants to pathogens," in *Biological Control of Plant Insects and Diseases,* ed. F. G. Maxwell and F. S. Harris (Jackson: University Press of Mississippi, 1974), pp. 403-461.

77. Swain, T. "Secondary compounds as protective agents," *Annual Rev. Plant Physiol.* 28 (1977): 479-501.

78. Kuć, J., "Phytoalexins," *Annual Rev. Phytopathol.* 10 (1972): 207-232.

79. Wood, R. K. S., and A. Graniti, eds., *Specificity in Plant Disease* (New York: Plenum Press, 1976).

80. Cochrane, V. W., "The role of plant residues in the etiology of root rot," *Phytopathology* 38 (1948): 185-96.

81. Toussoun, T. A., and Patrick, Z. A., "Effect of phytotoxic substances from decomposing plant residues on root rot of bean," *Phytopathology* 53 (1963): 265-70.

82. Patrick, Z. A., and L. W. Koch, "The adverse influence of phytotoxic substances from decomposing plant residues on resistance of tobacco to black root rot," *Canad. J. Bot.* 41 (1963): 747-58.

CHAPTER 4

1. Chitwood, B. G., "Introduction," in *An Introduction to Nematology,* ed. B. G. Chitwood and M. B. Chitwood (Baltimore: University Park Press, 1974).

2. Plinius Secundus, C., *Natural History,* trans. by H. Rackam, W. H. S. Jones, and D. E. Eichholz (Cambridge, Mass.: Harvard University Press, 1938-63).

3. Coles, W., *Adam in Eden: or Natures Paradise* (London: N. Brooke, 1957).

4. Tyler, J., "Proceedings of the root-knot nematode conference held at Atlanta, Georgia, February 4, 1938," *Plant Dis. Reptr. Suppl.* 109 (1938): 133-51.

5. Steiner, G., "Nematode parasitic on and associated with roots of marigold *(Tagetes hybrida),*" *Proc. Biol. Soc. Wash.* 54 (1941): 31-34.

6. Oostenbrink, M., M. Kuiper, and J. J. s'Jacobs, "*Tagetes* als Feindpflanzen von *Pratylenchus* Arten," *Nematologica Suppl.* 2 (1957): 424-33.

7. Uhlenbroek, J. H., and J. D. Bijloo "Investigations on nematicides.

I. Isolation and structure of a nematicidal principle occurring in *Tagetes* roots," *Rec. Trav. Chim. Pays-Bas* 77 (1958): 1004-1009.

8. Visser, T., and M. K. Vythilingam, "The effect of marigolds and some other crops on the *Pratylenchus* and *Meloidogyne* populations in tea soil," *Tea Quart.* 30 (1959): 30-38.

9. Omidvar, A. M., "On the effects of root diffusate from *Tagetes* spp. on *Heterodera rostochiensis*," *Nematologica* 6 (1961): 123-29.

10. Omidvar, A. M., "The nematicidal effects of *Tagetes* spp. on the final population of *Heterodera rostochiensis* Woll.," *Nematologica* 7 (1962): 62-64.

11. Hesling, J. J., K. Pawelska, and A. M. Shepherd, "The response of potato root eelworm, *Heterodera rostochiensis* Woll. and beet eelworm, *H. schactii* Schimidt to root diffusates of some grasses, cereals and of *Tagetes minuta*." *Nematologica* 6 (1961): 207-213.

12. Daulton, R. A. C., and R. F. Curtis, "The effects of *Tagetes* spp. on *Meloidogyne javanica* in southern Rhodesia," *Nematologica* 9 (1963): 357-62.

13. Daulton, R. A. C., "The behavior and control of root-knot nematode *Meloidogyne javanica* in tobacco as influenced by crop rotation and soil fumigation practices," paper delivered at Third World Tobacco Scientific Congress Proc., Salisbury, Southern Rhodesia, 1963.

14. Good, J. M., N. A. Minton, and C. A. Jaworski, "Relative susceptibility of selected cover crops and coastal bermudagrass to plant nematodes," *Phytopathology* 55 (1965): 1026-30.

15. Hackney, R. W. and O. J. Dickerson, "Marigold, castor bean and crysanthemum as controls of *Meloidogyne incognita* and *Pratylenchus alleni*," *J. Nematol.* 7 (1975): 84-90.

16. Khan, A. M., "Studies on Plant Parasitic Nematodes Associated with Vegetable Crops in Uttar Pradesh," Final Technical Report, Aligarh Muslim University, Aligarh, India, 1969.

17. Alam, M. M., S. K. Saxena, and A. M. Khan, "Influence of interculture of marigold and margosa with some vegetable crops on plant growth and nematode population," *Acta Bot. Indica* 5 (1977): 33-39.

18. Alam, M. M., A. Masood, and S. I. Husain, "Effect of margosa and marigold root-exudates on mortality and larval hatch of certain nematodes," *Indian J. Exp. Biol.* 13 (1975): 412-14.

19. Miller, P. M., and J. F. Ahrens, "Influence of growing marigolds, weeds, two cover crops and fumigation on subsequent populations of parasitic nematodes and plant growth," *Plant Dis. Rep.* 53 (1969): 642-46.

20. Winoto-Suatmadji, R., "Studies on the Effect of *Tagetes* spp. on Plant Parasitic Nematodes" (Wageningen, Netherlands: Veenman and Zonen N.V., 1969).

21. McBeth, C. W., and A. L. Taylor, "Immune and resistant cover crops valuable in root-knot infested peach orchards," *Amer. Soc. Hort. Sci. Proc.* 45 (1944): 158-66.

22. Oschse, J. J., and W. S. Brewton, "Preliminary report on *Crota-*

laria versus nematodes," *Florida Hort. Soc. Proc.* 67 (1954): 218-19.

23. Endo, B. Y., "Responses of root-lesion nematodes, *Pratylenchus brachyurus* and *P. zeae,* to various plants and soil types," *Phytopathology* 49 (1959): 417-21.

24. Hesling, J. J., and H. R. Wallace, "Susceptibility of varieties of crysanthemum to infestation by *Aphelenchoides ritzemabosi* (Schwartz)," *Nematologica* 5 (1960): 297-302.

25. Wallace, H. R., "The nature of resistance in crysanthemum varieties to *Aphelenchoides ritzemabosi,*" *Nematologica* 6 (1961): 49-58.

26. Rohde, R. A., and W. R. Jenkens, "Basis for resistance of *Asparagus officinalis* var. *altilis* L. to the stubby root nematode *Trichodorus christei* Allen, 1957," *Bull. Univ. Md. Agric. Sta. A.* 97 (1958): 19.

27. Husain, S. I., and A. Masood, "Effect of some plant extracts on larval hatching of *Meloidogyne incognita* (Jofoid and White) Chitwood," *Acta Bot. Indica* 3 (1975): 142-46.

CHAPTER 5

1. Wigglesworth, V. B., "Factors controlling moulting and metamorphosis in an insect," *Nature* 133 (1934): 725-26.

2. Richards, O. W., and R. G. Davies, *Imms' Outlines of Entomology,* 6th ed. (London: Chapman and Hall, 1978).

3. Eisner, T., and E. O. Wilson, *The Insects* (San Francisco: Freeman, 1977), pp. 1-15.

4. Dethier, V. G., "Some general considerations of insects' responses to the chemicals in food plants," in *Control of Insect Behavior by Natural Products,* ed. D. L. Wood, R. M. Silverstein, and M. Nakajima (New York: Academic Press, 1970), pp. 21-28.

5. Dethier V. G., L. B. Browne, and C. N. Smith, "The designation of chemicals in terms of the responses they elicit from insects," *J. Econ. Ent.* 53 (1960): 134-36.

6. Hedin, P. A., J. N. Jenkins, and F. G. Maxwell, "Behavioral and developmental factors affecting host plant resistance to insects," in *Host Plant Resistance to Pests,* ed. P. A. Hedin (Washington, D.C.: American Chemical Society, 1977), pp. 231-75.

7. Verschaffelt, E., "The cause determining the selection of food in some herbivorous insects," *Proc. Acad. Sci., Amsterdam* 13 (1910): 536-42.

8. McIndoo, N. E., "The olfactory sense of lepidopterous larvae," *Ann. Entomol. Soc. Amer.* 12 (1919): 65-84.

9. Dethier, V. G., "Gustation and olfaction in lepidopterous larvae," *Biol. Bull.* 72 (1937): 7-23.

10. Thorsteinson, A. J., "The chemotactic responses that determine host specificity in an oligophagous insect (*Plutella maculipennis* Curt.) Lepidoptera," *Canad. J. Zool.* 31 (1953): 52-72.

11. Yamamoto, I., and R. Yamamoto, "Host attractants for the rice weevil and the cheese mite," in *Control of Insect Behavior by Natural*

Products, ed. D. L. Wood, R. M. Silverstein, and M. Nakajima (New York: Academic Press, 1970), pp. 331-45.

12. Saito, T., and K. Munakata, "Insect attractants of vegetable origin, with special reference to the rice stem borer and fruit-piercing moths," in *Control of Insect Behavior by Natural Products,* ed. D. L. Wood, R. M. Silverstein, and M. Nakajima (New York: Academic Press, 1970), pp. 225-36.

13. Tumlinson, J. H., et al., "The boll weevil sex attractant," in *Chemicals Controlling Insect Behavior,* ed. M. Beroza (New York: Academic Press, 1970), pp. 41-59.

14. Matsumoto, Y., "Volatile organic sulfur compounds as insect attractants with special reference to host selection," in *Control of Insect Behavior by Natural Products,* ed. D. L. Wood, R. M. Silverstein, and M. Nakajima (New York: Academic Press, 1970), pp. 133-60.

15. Wood, D. L., "Pheromones of bark beetles," in *Control of Insect Behavior by Natural Products,* ed. D. L. Wood, R. M. Silverstein, and M. Nakajima (New York: Academic Press, 1970), pp. 301-16.

16. Rudinsky, J. A., "Various host-insect interrelations in host-finding and colonization behavior of bark beetles on coniferous trees," in *The Host-Plant in Relation to Insect Behaviour and Reproduction,* ed. T. Jermy, Symp. Biol. Hung. 16 (New York: Plenum Press, 1976), pp. 229-35.

17. Sutherland, O. R. W., "The attraction of the newly hatched coding moth *(Laspeyresia pomonella)* larvae to apple," *Ent. Exp. and Appl.* 15 (1972): 481-87.

18. Sutherland, O. R. W., and R. F. N. Hutchins, "α-Farnesene, a natural attractant for codling moth larvae," *Nature* 239 (1972): 170-71.

19. Page, C. R., III, and J. T. Barber, "Interactions between mosquito larvae and mucilaginous plant seeds II. Chemical attraction of larvae to seeds," *Mosquito News* 35 (1975): 47-54.

20. Wilde, J. de, "The olfactory component in host-plant selection in the adult Colorado beetle *(Leptinotarsa decemlineata* Say)," in *The Host-Plant in Relation to Insect Behaviour and Reproduction,* ed. T. Jermy, Symp. Biol. Hung. 16 (New York: Plenum Press, 1976), pp. 291-300.

21. Hawkes, C., and T. H. Coaker, "Behavioural responses to host-plant odours in adult cabbage root fly *(Erioischia brassicae* Bouché)," in *The Host-Plant in Relation to Insect Behaviour and Reproduction,* ed. T. Jermy, Symp. Biol. Hung. 16 (New York: Plenum Press, 1976), pp. 85-89.

22. Pettersson, J., "Ethology of *Dasyneura brassicae* Winn. (Dipt; Cecidomyidae). I. Laboratory studies of olfactory reactions to the host-plant," in *The Host-Plant in Relation to Behaviour and Reproduction,* ed. T. Jermy, Symp. Biol. Hung. 16 (New York: Plenum Press, 1976), pp. 203-208.

23. Sáringer, Gy., "Oviposition behaviour of *Ceutorrhynchus maculaalba* Herbst. (Col.: Curculionidae)," in *The Host-Plant in Relation to Behaviour and Reproduction,* ed. T. Jermy, Symp. Biol. Hung. 16 (New

NOTES

York: Plenum Press, 1976), pp. 241-45.

24. Marshall, D. L., A. J. Beattie, and W. E. Bollenbacher, "Evidence for diglycerides as attractants in an ant-seed interaction," *J. Chem. Ecol.* 5 (1979): 335-44.

25. Dethier, V. G., "The importance of stimulus patterns for host-plant recognition and acceptance," in *The Host-Plant in Relation to Insect Behaviour and Reproduction,* ed. T. Jermy, Symp. Biol. Hung. 16 (New York: Plenum Press, 1976), pp. 67-70.

26. Acree, F., "The chromatography of gyptol and gyptyl ester," *J. Econ. Ent.* 47 (1954): 321-26.

27. Gary, N. E., "Pheromones of the honey bee, *Apis mellifera* L.," in *Control of Insect Behavior by Natural Products,* eds. D. L. Wood, R. M. Silverstein, and M. Nakajima (New York: Academic Press, 1970), pp. 29-50.

28. Birch, M. C., ed., *Pheromones.* (Amsterdam: North-Holland, 1974).

29. Méry, F., *Animal Languages,* trans. Michael Ross (Farnborough, Hampshire, Eng.: Saxon House, Westmead, 1975).

30. Shorey, H. H., *Animal Communication by Pheromones* (New York: Academic Press, 1976).

31. Müller-Schwarze, D., and M. M. Mozell, eds., *Chemical Signals in Vertebrates* (New York: Plenum Press, 1977).

32. Ishii, S., "Aggregation of the German cockroach, *Blattella germanica* (L.)," in *Control of Insect Behavior by Natural Products,* ed. D. L. Wood, R. M. Silverstein, and M. Nakajima (New York: Academic Press, 1970), pp. 93-109.

33. Grevillius, A. Y., "Zur Kenntnis der Biologie des Goldafters (*Euproctis chrysorrhoea* [L.])," *Botan. Centr. Bieheft.* 18 (1905): 222-322.

34. Dethier, V. C., "Chemical factors determining choice of food plants by *Papilio* larvae," *Amer. Naturalist* 75 (1941): 61-73.

35. Fraenkel, G. S., "The raison d'être of secondary plant substances," *Science* 129 (1959): 1466-70.

36. Ehrlich, P. R., and P. H. Raven, "Butterflies and plants: a study in coevolution," *Evolution* 18 (1964): 586-608.

37. Chambliss, O. L., and C. M. Jones, "Cucurbitacins: Specific insect attractants in Cucurbitaceae," *Science* 153 (1966): 1392-93.

38. Rees, C. J. C., "Chemoreceptor specificity associated with choice of feeding site by the beetle, *Chrysolina brunsvicensis* on its foodplant, *Hypericum hirsutum,*" *Ent. Exp. and Appl.* 12 (1969): 565-83.

39. Bernays, E. A., and R. F. Chapman, "Plant chemistry and acridoid feeding behaviour," in *Biochemical Aspects of Plant and Animal Coevolution,* ed. J. B. Harborne (London: Academic Press, 1978), pp. 99-141.

40. Wearing, C. H., and R. F. N. Hutchins, "α-Farnesene, a naturally occurring oviposition stimulant for the codling moth *(Laspeyresia pomonella),*" *J. Insect Physiol.* 19 (1973): 1251-56.

41. Robert, P. Ch., "Inhibitory action of chestnut-leaf extracts (*Casta-*

nea sativa Mill.) on oviposition and oogenesis of the sugar beet moth (*Scrobipalpa ocellatella* Boyd.; Lepidoptera, Gelechiidae)," in *The Host-Plant in Relation to Insect Behaviour and Reproduction,* ed. T. Jermy, Symp. Biol. Hung. 16 (New York: Plenum Press, 1976), pp. 223-27.

42. Deseo, K. V., "The oviposition of the Indian meal moth *(Plodia interpunctella* HBN., Lep., Phyticidae) influenced by olfactory stimuli and antennectomy," in *The Host-Plant in Relation to Insect Behaviour and Reproduction,* ed. T. Jermy, Symp. Biol. Hung. 16 (New York: Plenum Press, 1976), pp. 61-65.

43. Huignard, J., "Interactions between the host-plant and mating upon the reproductive activity of *Acanthoscelides obtectus* females (Coleoptera, Bruchidae)," in *The Host-Plant in Relation to Insect Behaviour and Reproduction,* ed. T. Jermy, Symp. Biol. Hung. 16 (New York: Plenum Press, 1976), pp. 101-108.

44. Norris, D. M., "Physico-chemical aspects of the effects of certain phytochemicals on insect gustation," in *The Host-Plant in Relation to Insect Behaviour and Reproduction,* ed. T. Jermy, Symp. Biol. Hung. 16 (New York: Plenum Press, 1976), pp. 197-201.

45. Feeny, P., "Biochemical coevolution between plants and their insect herbivores," in *Coevolution of Animals and Plants,* ed. L. E. Gilbert and P. H. Raven (Austin: University of Texas Press, 1975), pp. 3-19.

46. Beroza, M., "Attractants and repellents for insect pest control," in *Pest Control Strategies for the Future,* ed. Agricultural Board, Division of Biology and Agriculture, National Academy of Sciences (Washington, D.C.: 1972), pp. 226-53.

47. Alexander, P., and D. H. R. Barton, "The excretion of ethylquinone by the flour beetle," *Biochem. J.* 37 (1943): 463-65.

48. Loconti, J. D., and L. M. Roth, "Composition of the odorous secretion of *Tribolium castaneum,*" *Ann. Ent. Soc. Amer.* 46 (1953): 281-89.

49. Jacobson, M., "Chemical insect attractants and repellents," *Annu. Rev. Ent.* 11 (1966): 403-22.

50. Klun, J. A., C. L. Tipton, and T. A. Brindley, "2,4-Dihydroxy-7-methoxy-1,4-benzoxazine-3-one (DIMBOA), an active agent in the resistance of maize to the European corn borer," *J. Econ. Ent.* 60 (1967): 1529-33.

51. Klun, J. A., and J. Robinson, "The concentration of two 1,4-benzoxazinones in dent corn at various stages of development of the plant and its relation to resistance of the host plant to the European corn borer," *J. Econ. Ent.* 62 (1969): 214-20.

52. Levin, D. A., "The role of trichomes in plant defense," *Quart. Rev. Biol.* 48 (1973): 3-15.

53. Penfold, A. R., and F. R. Morrison, "Some Australian essential oils in insecticides and repellents," *Soap, Perfumery Cosmetics* 25 (1952): 933-34.

54. Maxwell, F. G., J. N. Jenkins, and J. C. Keller, "A boll weevil

repellent from the volatile substance of cotton," *J. Econ. Ent.* 56 (1963): 894-95.

55. Eisner, T., "Catnip: its raison d'être," *Science* 146 (1964): 1318-20.

56. Hsiao, T. H., "Chemical and behavioral factors influencing food selection of *Leptinotarsa* beetle," in *The Host-Plant in Relation to Insect Behaviour and Reproduction,* ed. T. Jermy, Symp. Biol. Hung. 16 (New York: Plenum Press, 1976), pp. 95-99.

57. Karasev, V. S., "The role of volatile oil composition for trunk pest resistance in coniferous plants. Experiments on lumber," in *The Host-Plant in Relation to Insect Behaviour and Reproduction,* ed. T. Jermy, Symp. Biol. Hung. 16 (New York: Plenum Press, 1976), pp. 115-19.

58. Norris, D. M., "Role of repellents and deterrents in feeding of *Scolytus multistriatus,"* in *Host Plant Resistance to Pests,* ed. P. A. Hedin (Washington, D.C.: American Chemical Society, 1977), pp. 215-30.

59. Maxwell, F. G., "Biologically active substances in cotton and related plants that affect boll weevil behavior and development," in *Insect-Plant Interactions,* ed. United States National Committee for International Biological Program (Washington, D.C.: National Academy of Sciences, 1969), pp. 45-49.

60. Lichtenstein, E. P., F. M. Strong, and D. G. Morgan, "Identification of 2-phenylethylisothiocyanate as an insecticide occurring naturally in the edible part of turnips," *J. Agr. Food Chem.* 10 (1962): 30-33.

61. Lichtenstein, E. P., and J. E. Casida, "Myristicin, an insecticide. and synergist occurring naturally in the edible parts of parsnips," *J. Agr. Food Chem.* 11 (1963): 410-15.

62. Merz, E., "Pflanzen und Raupen. Über einige Prinzipien der Futter-wahl bei Grossschmetter-lingsraupen," *Biol. Zentr.* 78 (1959): 152-88.

63. Kogan, M., "Plant resistance in pest management," in *Introduction to Insect Pest Management,* ed. R. Metcalf and W. H. Luckman (New York: Wiley and Sons, 1975), pp. 103-46.

64. Hardee, D. D., and T. B. Davich, "A feeding deterrent for the boll weevil, *Anthonomus grandis,* from tung meal," *J. Econ. Ent.* 59 (1966): 1267-70.

65. Munakata, K., "Insect antifeedants in plants," in *Control of Insect Behavior by Natural Products,* ed. D. L. Wood, R. M. Silverstein, and M. Nakajima (New York: Academic Press, 1970), pp. 179-87.

66. Soo Hoo, C. F., and G. Fraenkel, "The resistance of ferns to the feeding of *Prodenia eridania* larvae," *Ann. Entomol. Soc. Amer.* 57 (1964): 788-90.

67. Reed, G. L., T. A. Brindley, and W. G. Showers, "Influence of resistant corn leaf tissue on the biology of the European corn borer," *Ann. Entomol. Soc. Amer.* 65 (1972): 658-62.

68. Chippendale, G. M., and G. P. V. Reddy, "Dietary sterols: Role in larval feeding behaviour of the southwestern corn borer, *Diatraea grandiosella,"* *Experientia* 28 (1974): 485-86.

69. Ehrlich, P. R., "Coevolution and the biology of communities,"

in *Biochemical Coevolution,* ed. K. L. Chambers, Biol. Colloquium #29 (Corvallis: Oregon State University Press, 1970), pp. 1-11.

70. Bernays, E. A., and R. F. Chapman, "Experiments to determine the basis of food selection by *Chorthippus parallelus* (Zetterstedt) (Orthoptera: Acrididae) in the field," *J. Anim. Ecol.* 39 (1970): 761-76.

71. Bernays, E. A., and R. F. Chapman, "The role of food plants in the survival and development of *Chortoicetes terminifera* (Walker) under drought conditions," *Aust. J. Zool.* 21 (1973): 575-92.

72. Blaney, W. M., and R. F. Chapman, "The function of the maxillary palps of Acrididae (Orthoptera)," *Ent. Exp. and Appl.* 13 (1970): 363-76.

73. Bernays, E. A., et al., "The ability of *Locusta migratoria* L. to perceive plant surface waxes," in *The Host-Plant in Relation to Insect Behaviour and Reproduction,* ed. T. Jermy, Symp. Biol. Hung. 16 (New York: Plenum Press, 1976), pp. 35-40.

74. Rockwood, L. L., "Seasonal changes in the susceptibility of *Crescentia alata* leaves to the flea beetle, *Oedionychus* sp.," Ecology 55 (1974): 142-48.

75. Feeny, P. P., "Effect of oak leaf tannins on larval growth of the winter moth *Operophtera brumata,*" *J. Insect Physiol.* 14 (1968): 805-17.

76. Feeny, P. P., and H. Bostock, "Seasonal changes in the tannin content of oak leaves," *Phytochemistry* 7 (1968): 871-80.

77. Bernays, E. A., and R. F. Chapman, "Antifeedant properties of seedling grasses," in *The Host-Plant in Relation to Insect Behaviour and Reproduction,* ed. T. Jermy, Symp. Biol. Hung. 16 (New York: Plenum Press, 1976), pp. 41-46.

78. Barry, B. D., J. A. Burnside, and H. S. Myers, Cucumis *species resistance to striped cucumber beetle seedling feeding and bacterial wilt,* Agr. Res. Ser. NC-46 (1976), 33 pages.

79. Juneja, P. S., R. K. Gholson, R. L. Burton and K. J. Starks, "The chemical basis for greenbug resistance in small grains. I. Benzyl alcohol as a possible resistance factor," *Ann. Entomol. Soc. Amer.* 65 (1972): 961-64.

80. Bell, E. A., "Toxins in seeds," in *Biochemical Aspects of Plant and Animal Coevolution,* ed. J. B. Harborne (New York: Academic Press, 1978), pp. 143-61.

81. Burnett, W. C., Jr., S. B. Jones, Jr., and T. J. Mabry, "The role of sesquiterpene lactones in plant-animal coevolution," in *Biochemical Aspects of Plant and Animal Coevolution,* ed. J. B. Harborne (London: Academic Press, 1978), pp. 233-57.

82. Brower, L. P., "Plant poisons in a terrestrial food chain and implications for mimicry theory," in *Biochemical Coevolution,* ed. K. L. Chambers, 29th Biol. Colloquium (Corvallis: Oregon State University Press, 1970), pp. 69-82.

83. Brower, L. P., and S. C. Glazier, "Localization of heart poisons in the monarch butterfly," *Science* 188 (1975): 19-25.

84. Todd, G. W., A. Getahun, and D. C. Cress, "Resistance in barley to the greenbug, *Schizaphis graminum.* I. Toxicity of phenolic and flavonoid compounds and related substances," *Ann. Entomol. Soc. Amer.* 64 (1971): 718-22.

85. Erickson, J. M., and P. Feeny, "Sinigrin: a chemical barrier to the black swallowtail butterfly, *Papilio polyxenes,*" *Ecology* 55 (1974): 103-11.

86. Bailey, J. C., F. G. Maxwell, and J. N. Jenkins, "Boll weevil antibiosis studies with selected cotton lines utilizing egg-plantation techniques," *J. Econ. Ent.* 60 (1967): 1275-79.

87. Agarwal, R. A., and N. Krishnananda, "Preference to oviposition and antibiosis mechanism of jassids (*Amrasca devastans* Dist.) in cotton (*Gossypium* sp.)," in *The Host-Plant in Relation to Insect Behaviour and Reproduction,* ed. T. Jermy, Symp. Biol. Hung. 16 (New York: Plenum Press, 1976), pp. 13-22.

88. Lukefahr, M. J., and D. F. Martin, "Cotton-plant pigments as a source of resistance to the bollworm and tobacco budworm," *J. Econ. Ent.* 59 (1966): 176-79.

89. Reese, J. C., and S. D. Beck, "Effects of certain allelochemics on the growth and development of the black cutworm," in *The Host-Plant in Relation to Insect Behaviour and Reproduction,* ed. T. Jermy, Symp. Biol. Hung. 16 (New York: Plenum Press, 1976), pp. 217-21.

90. Reese, J. C., "The effects of plant biochemicals on insect growth and nutritional physiology," in *Host-Plant Resistance to Pests,* ed. P. A. Hedin (Washington, D.C.: American Chemical Society, 1977), pp. 129-52.

91. Janzen, D. H., H. B. Juster, and E. A. Bell, "Toxicity of secondary compounds to the seed-eating larvae of the bruchid beetle *(Callosobruchus maculatus),*" *Phytochemistry* 16 (1977): 223-27.

92. Free, J. B., *Insect Pollination of Crops* (New York: Academic Press, 1970).

93. Huber, F., *New Observations on the Natural History of Bees,* 3d ed. (London: W. & C. Tait, and Longman, Hurst, Rees, Orme, and Brown, 1821).

94. Dobbs, A., "Concerning bees, and their method of gathering wax and honey," *Philos. Trans. Roy. Soc.* 46 (1750): 536-49.

95. Lepage, M., and R. Boch, "Pollen lipids attractive to honeybees," *Lipids* 3 (1968): 530-34.

96. Hopkins, C. Y., A. W. Jevans, and R. Bock, "Occurrence of octadeca-*trans*-2-*cis* 9, *cis*-12-trienoic acid in pollen attractive to the honey bee," *Canad. J. Biochem.* 47 (1969): 433-36.

97. Pijl, L. van der, and C. H. Dodson, *Orchid Flowers: Their Pollination and Evolution* (Coral Gables: University of Miami Press, 1966).

98. Dodson, C. H., and H. G. Hills, "Gas chromatography of orchid fragrances," *Amer. Orch. Soc. Bull.* 35 (1966): 720-25.

99. Dodson, C. H., R. L. Dressler, H. G. Hills, R. M. Adams and N. H. Norris, "Biologically active compounds in orchid fragrances," *Science*

164 (1969): 1243-49.

100. Dodson, C. H., "Coevolution of orchids and bees," in *Coevolution of Animals and Plants,* ed. L. E. Gilbert and P. H. Raven (Austin: University of Texas Press, 1975), pp. 91-99.

101. Brantjes, N. B. M., "Sensory responses to flowers in night-flying moths," in *The Pollination of Flowers by Insects,* ed. A. J. Richards (London: Academic Press for The Linnean Society of London, 1978), pp. 13-19.

102. Meeuse, B. J. D., "The physiology of some sapromyophilous flowers," in *The Pollination of Flowers by Insects,* ed. A. J. Richards (London: Academic Press for The Linnean Society of London, 1978), pp. 97-104.

CHAPTER 6

1. Williams, C. M., "Third-generation pesticides," *Scient. Amer.* 217(1) (1967): 13-17.

2. Feeny, P., "Biochemical coevolution between plants and their insect herbivores," in *Coevolution of Animals and Plants,* ed. L. E. Gilbert and P. H. Raven (Austin: University of Texas Press, 1975), pp. 3-19.

3. Horn, D. H. S., "The ecdysones," in *Naturally Occurring Insecticides,* ed. M. Jacobson and D. G. Crosby (New York: Marcel Dekker, 1971), pp. 333-459.

4. DeBach, P., *Biological Control by Natural Enemies* (London: Cambridge University Press, 1974).

5. Nelson, R. R., "Introduction," in *Breeding Plants for Disease Resistance—Concepts and Applications,* ed. R. R. Nelson (University Park: Pennsylvania State University Press, 1973), pp. 3-12.

6. Kingsbury, J. M., *Poisonous Plants of the United States and Canada* (Englewood Cliffs, N.J.: Prentice-Hall, 1964).

7. Bell, E. A., "Toxins in seeds," in *Biochemical Aspects of Plant and Animal Coevolution,* ed. J. B. Harborne (London: Academic Press, 1978), pp. 143-61.

8. Toms, G. C., and A. Western, "Phytohaemagglutinins," in *Chemotaxonomy of the Leguminosae,* ed. J. B. Harborne, J. B. Boulter, and B. L. Turner (London: Academic Press, 1971), pp. 367-456.

9. Janzen, D. H., H. B. Juster, and I. E. Liener, "Insecticidal action of the phytohemagglutinin in black beans on a bruchid beetle," *Science* 192 (1976): 795-96.

10. Janzen, D. H., H. B. Juster, and E. A. Bell, "Toxicity of secondary compounds to the seed-eating larvae of the bruchid beetle *Callosobruchus maculatus,*" *Phytochemistry* 16 (1977): 223-27.

11. Applebaum, S. W., and Y. Birk, "Natural mechanisms of resistance to insects in legume seeds," in *Insect and Mite Nutrition,* ed. J. G. Rodriguez (Amsterdam: North-Holland, 1972), pp. 629-36.

12. Patrick, Z. A., "The peach replant problem in Ontario. II. Toxic

substances from microbial decomposition products of peach root residues," *Canad. J. Bot.* 33 (1955): 461-86.

13. Seigler, D., "Determination of cyanolipids in seed oils of the Sapindaceae by means of their NMR spectra," *Phytochemistry* 13 (1974): 841-43.

14. Smolenski, S. J., H. Silinis, and N. R. Farnsworth, "Alkaloid screening. VII," *Lloydia* 38 (1975): 411-41.

15. Bell, E. A., "'Uncommon' amino acids in plants," *FEBS Letts.* 64 (1976): 29-35.

16. Lichtenstein, E. P., F. M. Strong, and D. G. Morgan, "Identification of 2-phenylethylisothiocyanate as an insecticide occurring naturally in the edible part of turnips," *J. Agr. Food Chem.* 10 (1962): 30-33.

17. Lichtenstein, E. P., and J. E. Casida, "Myristicin, an insecticide and synergist occurring naturally in the edible parts of parsnips," *J. Agr. Food Chem.* 11 (1963): 410-15.

18. Busbey, R. L., "Plants that help kill insects," in United States Department of Agriculture, *The Yearbook of Agriculture: Crops in Peace and War* (Washington, D.C.: Government Printing Office, 1950-51), pp. 765-71.

19. Schmeltz, I., "Nicotine and other tobacco alkaloids," in *Naturally Occurring Insecticides,* ed. M. Jacobson and D. G. Crosby (New York: Marcel Dekker, 1971), pp. 99-136.

20. Matsui, M., and I. Yamamoto, "Pyrethroids," in *Naturally Occurring Insecticides,* ed. M. Jacobson and D. G. Crosby (New York: Marcel Dekker, 1971), pp. 3-70.

21. Casida, J. E., ed., *Pyrethrum— The Natural Insecticide* (New York: Academic Press, 1973).

22. Fukami, H., and M. Nakajima, "Rotenone and the Rotenoids," in *Naturally Occurring Insecticides,* ed. M. Jacobson and D. G. Crosby (New York: Marcel Dekker, 1971), pp. 71-97.

23. Jacobson, M., "The unsaturated isobutyl-amides," in *Naturally Occurring Insecticides,* ed. M. Jacobson and D. G. Crosby (New York: Marcel Dekker, 1971), pp. 137-76.

24. Crosby, D. G., "Minor insecticides of plant origin," in *Naturally Occurring Insecticides,* ed. M. Jacobson and D. G. Crosby (New York: Marcel Dekker, 1971), pp. 177-239.

25. Penfold, A. R., and F. R. Morrison, "Some Australian essential oils in insecticides and repellents," *Soap, Perfumery Cosmetics:* 25 (1952): 933-34.

26. Painter, R. R., "Repellents," in *Pest Control—Biological, Physical and Selected Chemical Methods,* ed. W. W. Kilgore and R. L. Doutt (New York: Academic Press, 1967), pp. 267-85.

27. Beroza, M., "Current usage and some recent developments with insect attractants and repellents in the USDA," in *Chemicals Controlling Insect Behavior,* ed. M. Beroza (New York: Academic Press, 1970), pp. 145-63.

28. Metcalf, R. L., and R. A. Metcalf, "Attractants, repellents, and genetic control in pest management," in *Introduction to Insect Pest Management,* ed. R. L. Metcalf and W. H. Luckmann (New York: Wiley and Sons, 1975), pp. 275-306.

29. Woodrow, A. W., N. Green, H. Tucker, M. H. Schonhorst, and K. C. Hamilton, "Evaluation of chemicals as honey bee attractants and repellents," *J. Econ. Ent.* 58 (1965): 1094-1102.

30. Wright, D. P., Jr., "Antifeedants," in *Pest Control: Biological, Physical, and Selected Chemical Methods,* ed. W. W. Kilgore and R. L. Doutt (New York: Academic Press, 1967), pp. 287-93.

31. Munakata, K., "Insect antifeedants in plants," in *Control of Insect Behavior by Natural Products,* ed. D. L. Wood, R. M. Silverstein, and M. Nakajima (New York: Academic Press, 1970), pp. 179-87.

32. Capinera, J. L., and F. R. Stermitz, "Laboratory evaluation of zanthophylline as a feeding deterrent for range catepillar, migratory grasshopper, alfalfa weevil, and greenbug," *J. Chem. Ecol.* 5 (1979): 767-71.

33. Sáringer, Gy., "Oviposition behavior of *Ceutorrhynchus maculaalba* Herbst. (Col.: Curculionidae)," in *The Host-Plant in Relation to Insect Behaviour and Reproduction,* ed. T. Jermy, Symp. Biol. Hung. 16 (New York: Plenum Press, 1976), pp. 241-45.

34. Robert, P. Ch., "Inhibitory action of chestnut-leaf extracts (*Castanea sativa* Mill.) on oviposition and oogenesis of the sugar beet moth (*Scrobipalpa ocellatella* Boyd; Lepidoptera, Gelechiidae)," in *The Host-Plant in Relation to Insect Behavior and Reproduction,* ed. T. Jermy (New York: Plenum Press, 1976), pp. 223-27.

35. Beroza, M., "Control of the gypsy moth and other insects with behavior-controlling chemicals," in *Pest Management with Insect Attractants and Other Behavior-Controlling Chemicals,* ed. M. Beroza (Washington, D.C.: American Chemical Society, 1976), pp. 99-118.

36. Reynolds, H. T., P. L. Adkisson, and R. F. Smith, "Cotton insect pest management," in *Introduction to Insect Pest Management,* ed. R. L. Metcalf and W. H. Luckmann (New York: Wiley and Sons, 1975), pp. 379-443.

37. Hedin, P. A., R. C. Gueldner, and A. C. Thompson, "Utilization of the boll weevil pheromone for insect control," in *Pest Management with Insect Attractants and Other Behavior-Controlling Chemicals,* ed. M. Beroza (Washington, D.C.: American Chemical Society, 1976), pp. 30-52.

38. Cross, W. H., and H. C. Mitchell, "Mating behavior of the female boll weevil," *J. Econ. Ent.* 59 (1966): 1503-1507.

39. Hardee, D. D., O. H. Lindbig, and T. B. Davich, "Suppression of populations of boll weevils over a large area in west Texas with pheromone traps in 1969," *J. Econ. Ent.* 64 (1971): 928-33.

40. Shorey, H. H., L. K. Gaston, and R. S. Kaae, "Air-permeation with gossyplure for control of the pink bollworm," in *Pest Management with Insect Attractants and Other Behavior-Controlling Chemicals,* ed.

M. Beroza (Washington, D.C.: American Chemical Society, 1976), pp. 67-74.

41. Shorey, H. H., R. S. Kaae, and L. K. Gaston, "Sex pheromones of Lepidoptera. Development of a method for pheromonal control of *Pectinophora gossypiella* in cotton," *J. Econ. Ent.* 67 (1974): 347-50.

42. Klun, J. A., and G. A. Junk, "Iowa European corn borer sex pheromones: Isolation and identification of four C_{14} esters," *J. Chem. Ecol.* 3 (1977): 447-59.

43. Klun, J. A., S. Maini, O. L. Chapman, G. Lepone, and G. H. Lee, "Suppression of male European corn borer sex attraction and precopulatory reactions with (E)-9-tetradecenyl acetate," *J. Chem. Ecol.* 5 (1979): 345-52.

44. Metcalf, R. L., "Insecticides in pest management," in *Introduction to Insect Pest Management,* ed. R. L. Metcalf and W. H. Luckmann (New York: Wiley and Sons, 1975), pp. 235-73.

45. Madsen, H. F., and J. M. Vakenti, "Codling moth: Use of Codlemone(R)-baited traps and visual detection of entries to determine need of sprays," *Environ. Ent.* 2 (1973): 677-79.

46. Bierl, B. A., M. Beroza, and C. W. Collier, "Potent sex attractant of the gypsy moth: Its isolation, identification, and synthesis," *Science* 170 (1970): 87-89.

47. Wigglesworth, V. B., *Insect Hormones* (San Francisco: Freeman, 1970).

48. Bowers, W. S., "Juvenile hormones," in *Naturally Occurring Insecticides,* ed. M. Jacobson and D. G. Crosby (New York: Marcel Dekker, 1971), pp. 307-32.

49. Robbins, W. E., "Hormonal chemicals for invertebrate pest control," in *Pest Control Strategies for the Future,* ed. Agricultural Board, Division of Biology and Agriculture, National Research Council (Washington, D.C.: National Academy of Sciences, 1972), pp. 172-96.

CHAPTER 7

1. Lanham, U., *The Insects* (New York: Columbia University Press, 1964).

2. Chaplin, C. E., L. P. Stoltz, and J. G. Rodriguez, "The inheritance of resistance to the two-spotted mite, *Tetranychus urticae* Koch, in strawberries," *Proc. Amer. Soc. Hort. Sci.* 92 (1968): 376-80.

3. Rodriguez, J. G., Z. T. Dabrowski, L. P. Stoltz, C. E. Chaplin and W. O. Smith, Jr., "Studies on resistance of strawberries to mites. 2. Preference and nonpreference responses of *Tetranychus urticae* and *T. turkestani* to water-soluble extracts of foliage," *J. Econ. Ent.* 64 (1971): 383-87.

4. Dabrowski, Z. T., and J. G. Rodriguez, "Studies on resistance of strawberries to mites. 3. Preference and nonpreference responses of

Tetranychus urticae and *T. turkestani* to essential oils of foliage," *J. Econ. Ent.* 64 (1971): 387-91.

5. Yamamoto, I., and R. Yamamoto, "Host attractants for the rice weevil and the cheese mite," in *Control of Insect Behavior by Natural Products,* ed. D. L. Wood, R. M. Silverstein, and M. Nakajima (New York: Academic Press, 1970), pp. 331-45.

6. Faegri, K. and K. van der Pijl, *The Principles of Pollination Ecology,* 2d ed. (New York: Pergamon Press, 1971).

7. Dabrowski, Z. T., "Some new aspects of host-plant relation to behavior and reproduction of spider mites (Acarina: Tetranychidae)," in *The Host-Plant in Relation to Insect Behaviour and Reproduction,* ed. T. Jermy, Symp. Biol. Hung. 16 (New York: Plenum Press, 1976), pp. 55-60.

8. Jones, D. A., "Selective eating of the acyanogenic form of the plant *Lotus corniculatus* L. by various animals," *Nature* 193 (1962): 1109-10.

9. Crawford-Sidebotham, T. J., "The role of slugs and snails in the maintenance of the cyanogenesis polymorphisms of *Lotus corniculatus* and *Trifolium repens,"* *Heredity* 28 (1972): 405-11.

10. Ellis, W. M., R. J. Keymer, and D. A. Jones, "The defensive function of cyanogenesis in natural populations," *Experientia* 33 (1977): 309-11.

11. Angseesing, J. P. A., and W. J. Angseesing, "Field observations on the cyanogenesis polymorphism of *Trifolium repens,"* *Heredity* 31 (1973): 276-82.

12. Dement, W. A., and H. A. Mooney, "Seasonal variation in the production of tannins and cyanogenic glucosides in the chaparral shrub, *Heteromeles arbutifolia,"* *Oecologia* 15 (1974): 65-76.

13. Sherbrooke, W. C., "Differential acceptance of toxic jojoba seed *(Simmondsia chinensis)* by four Sonoran Desert heteromyid rodents," *Ecology* 57 (1976): 596-602.

14. Rosenzweig, M. L., and J. Winakur, "Population ecology of desert rodent communities: habitats and environmental complexity," *Ecology* 50 (1969): 558-72.

15. Cooper-Driver, G. A., and T. Swain, "Cyanogenic polymorphism in bracken in relation to herbivore predation," *Nature* 260 (1976): 604.

16. Burnett, W. C., Jr., S. B. Jones, Jr., and T. J. Mabry, "The role of sesquiterpene lactones in plant-animal coevolution," in *Biochemical Aspects of Plant and Animal Coevolution,* ed. J. B. Harborne (New York: Academic Press, 1978), pp. 233-57.

17. Dimock, E. J., II, R. R. Silen, and V. E. Allen, "Genetic resistance in Douglas-fir to damage by snowshoe hare and black-tailed deer," *For. Sci.* 22 (1976): 106-21.

18. Radwan, M. A., and G. L. Crouch, "Selected chemical constituents and deer browsing preference of Douglas fir," *J. Chem. Ecol.* 4 (1978): 675-83.

19. Roe, R., and B. E. Mottershead, "Palatability of *Phalaris arundinacea* L.," *Nature* 193 (1962): 255-56.

20. Williams, M., R. F. Barnes and J. M. Cassady, "Characterization of alkaloids in palatable and unpalatable clones of *Phalaris arundinacea* L.," *Crop Sci.* 11 (1971): 213-17.

21. Atsatt, P., and D. O'Dowd, "Mutual aid among plants," *Horticulture* 56 (1978): 22-31.

22. Kingsbury, J. M., *Poisonous Plants of the United States and Canada* (Englewood Cliffs, N.J.: Prentice-Hall, 1964).

23. Rice, E. L., *Allelopathy* (New York: Academic Press, 1974).

24. Rodriguez, J. G., "Inhibition of acarid mite development by fatty acids," in *Insect and Mite Nutrition,* ed. J. G. Rodriguez (Amsterdam: North-Holland, 1972), pp. 637-50.

25. Bate-Smith, E. C., "Plant phenolics in foods," in *The Pharmacology of Plant Phenolics,* ed. J. W. Fairbairn (New York: Academic Press, 1959), pp. 133-49.

26. Rayudu, G. V. N., P. Kadirvel, P. Vohra, and F. H. Kratzer, "Toxicity of tannic acid and its metabolites for chickens," *Poultry Sci.* 49 (1970): 957-60.

27. Vohra, P., F. H. Kratzer, and M. A. Joslyn, "The growth depressing and toxic effects of tannins to chicks," *Poultry Sci.* 45 (1966): 135.

28. Joslyn, M. A., and Z. Glick, "Comparative effects of gallotannic acid and related phenolics on the growth of rats," *J. Nutr.* 98 (1969): 119-26.

29. Peaslee, M. H., and F. A. Einhellig, "Tannic acid-induced alterations in mouse growth and pituitary melanocyte-stimulating hormone activity," *Toxicol. & Applied Pharmacol.* 25 (1973): 507-14.

30. Peaslee, M. H., and F. A. Einhellig, "Reduced fecundity in mice on tannic acid diet," *Comp. Gen. Pharmacol.* 4 (1973): 393-97.

31. Korpássy, B., "The hepatocarcinogenicity of tannic acid," *Cancer Res.* 19 (1959): 501-504.

32. Morton, J. F., "Economic botany in epidemiology," *Econ. Bot.* 32 (1978): 111-18.

33. Hungate, R. E., *The Rumen and Its Microbes* (New York: Academic Press, 1966).

34. Smart, W. W. G., Jr., T. A. Bell, N. W. Stanley, and W. A. Cope, "Inhibition of rumen cellulase by extract of sericea forage," *J. Dairy Sci.* 44 (1961): 1945-46.

35. Nagy, J. G., G. Vidacs, and G. M. Ward, "Separation of the essential oils of *Artemisia* spp. by gas chromatography and the effects of the oils on bacteria (Abstract)," *J. Colo.-Wyoming Acad. Science* 5 (1964): 41-42.

36. Sidhu, K. S., and W. H. Pfander, "Metabolic inhibitor(s) in orchard grass (*Dactylis glomerata* L.)," *J. Dairy Sci.* 51 (1968): 1042-45.

37. Harris, H. B., D. G. Cummins, and R. E. Burns, "Tannin content

and digestibility of sorghum grain as influenced by bagging," *Agron. J.* 62 (1970): 633-35.

38. Cummins, D. G., "Relationships between tannin content and forage digestibility in sorghum," *Agron. J.* 63 (1971): 500-502.

39. Shorey, H. H., *Animal Communication by Pheromones* (New York: Academic Press, 1976).

40. Mykytowycz, R., "Odor in the spacing behavior of mammals," in *Pheromones,* ed. M. C. Birch (Amsterdam: North-Holland, 1974), pp. 327-43.

41. Stoddart, D. M., "The role of odor in the social biology of small mammals," in *Pheromones,* ed. M. C. Birch (Amsterdam: North-Holland, 1974), pp. 297-315.

42. Ropartz, Ph., "Chemical signals in agonistic and social behavior of rodents," in *Chemical Signals in Vertebrates,* ed. D. Müller-Schwarze and M. M. Mozell (New York: Plenum Press, 1977), pp. 169-84.

43. Bowers, J. M., and B. K. Alexander, "Mice: individual recognition by olfactory cues," *Science* 158 (1967): 1208-1210.

44. Bronson, F. H., "Pheromonal influences on reproductive activities in rodents," in *Pheromones,* ed. M. C. Birch (Amsterdam: North-Holland, 1974), pp. 344-65.

45. Calhoun, J. B., *The Ecology and Sociology of the Norway Rat,* USPHS Publ. No. 1008 (Washington, D.C.: United States Government Printing Office, 1962).

46. Carr, W. J., L. S. Loeb, and M. L. Dissinger, "Responses of rats to sex odors," *J. Comp. Physiol. Psychol.* 59 (1965): 370-77.

47. Stacewicz-Sapuntzakis, M., and A. M. Gawienowski, "Rat olfactory response to aliphatic acetates," *J. Chem. Ecol.* 3 (1977): 411-17.

48. Gawienowski, A. M., D. A. Keedy, and M. Stacewicz-Sapuntzakis, "Bioassay apparatus for rodent olfactory preferences under laboratory and field conditions," *J. Chem. Ecol.* 5 (1979): 595-601.

49. Mykytowycz, R., "Olfaction in relation to reproduction in domestic animals," in *Chemical Signals in Vertebrates,* ed. D. Müller-Schwarze and M. M. Mozell (New York: Plenum Press, 1977), pp. 207-24.

50. Estes, R. D., "The role of the vomeronasal organ in mammalian reproduction," *Mammalia* 36 (1972): 315-41.

51. Donovan, C. A., "Some clinical observations on sexual attraction and deterrence in dogs and cattle," *Vet. Med. Small Anim. Clin.* 62 (1967): 1047-51.

52. Fletcher, I. C., and D. R. Lindsay, "Sensory involvement in the mating behavior of domestic sheep," *Anim. Behav.* 16 (1968): 410-14.

53. Müller-Schwarze, D., "Application of pheromones in mammals," in *Pheromones,* ed. M. C. Birch (Amsterdam: North-Holland, 1974), pp. 452-54.

54. Müller-Schwarze, D., "Responses of young black-tailed deer to predator odors," *J. Mammal.* 53 (1972): 393-94.

55. Müller-Schwarze, D., "Phermones in black-tailed deer," *Anim. Behav.* 19 (1971): 141-52.

CHAPTER 8

1. Rice, E. L., *Allelopathy* (New York: Academic Press, 1974).
2. Jacobson, M. "Chemical insect attractants and repellents," *Annu. Rev. Ent.* 11 (1966): 403-22.
3. Karasev, V. S., "The role of volatile oil composition for trunk pest resistance in coniferous plants. Experiments on lumber," in *The Host-Plant in Relation to Insect Behaviour and Reproduction,* ed. T. Jermy, Symp. Biol. Hung. 16 (New York: Plenum Press, 1976), pp. 115-19.
4. Rudinsky, J. A., "Various host-insect inter-relations in host-finding and colonization behavior of bark beetles on coniferous trees," in *The Host-Plant in Relation to Behaviour and Reproduction,* ed. T. Jermy, Symp. Biol. Hung. 16 (New York: Plenum Press, 1976), pp. 229-35.
5. Dodson, C. H., R. L. Dressler, H. G. Hills, R. M. Adams, and N. H. Norris, "Biologically active compounds in orchid fragrances," *Science* 164 (1969): 1243-49.
6. Todd, G. W., A. Getahun, and D. C. Cress, "Resistance in barley to the greenbug, *Schizaphis graminum.* I. Toxicity of phenolic and flavonoid compounds and related substances," *Ann. Entomol. Soc. Amer.* 64 (1971): 718-22.
7. Feeny, P. P., and H. Bostock, "Seasonal changes in the tannin content of oak leaves," *Phytochemistry* 7 (1968): 871-80.
8. Vohra, P., F. H. Kratzer, and M. A. Joslyn, "The growth depressing and toxic effects of tannins to chicks," *Poultry Sci.* 45 (1966): 135.
9. Joslyn, M. A., and Z. Glick, "Comparative effects of gallotannic acid and related phenolics on the growth of rats," *J. Nutr.* 98 (1960): 119-26.
10. Peaslee, M. H., and F. A. Einhellig, "Tannic acid-induced alterations in mouse growth and pituitary melanocyte-stimulating hormone activity," *Toxicol. & Applied Pharmacol.* 25 (1973): 507-14.
11. Feeny, P. P., "Effect of oak leaf tannins on larval growth of the winter moth *Operophtera brumata,"J. Insect Physiol.* 14 (1968): 805-17.
12. Radwan, M. A., and G. L. Crouch, "Selected chemical constituents and deer browsing preference of Douglas fir," *J. Chem. Ecol.* 4 (1978): 675-83.
13. Dabrowski, Z. T., "Some new aspects of host-plant relation to behaviour and reproduction of spider mites (Acarina: Tetranychidae)," in *The Host-Plant in Relation to Insect Behaviour and Reproduction,* ed. T. Jermy (New York: Plenum Press, 1976), pp. 55-60.
14. Rodriguez, J. G., "Inhibition of acarid mite development by fatty acids," in *Insect and Mite Nutrition,* ed J. G. Rodriguez (Amsterdam: North-Holland, 1972), pp. 637-50.

15. Mykytowycz, R., "Odor in the spacing behavior of mammals," in *Pheromones,* ed. M. C. Birch (Amsterdam: North-Holland, 1974), pp. 327-43.

16. Bronson, F. H., "Pheromonal influences on reproductive activities in rodents," in *Pheromones,* ed. M. C. Birch (Amsterdam: North-Holland, 1974), pp. 344-65.

17. Stacewicz-Sapuntzakis, M., and A. M. Gawienowski, "Rat olfactory response to aliphatic acetates," *J. Chem. Ecol.* 3 (1977): 411-17.

18. Mykytowycz, R. "Olfaction in relation to reproduction in domestic animals," in *Chemical Signals in Vertebrates,* ed. D. Müller-Schwarze and M. M. Mozell (New York: Plenum Press, 1977), pp. 207-24.

19. Robinson, E. L., "Effect of weed species and placement on seed cotton yields," *Weed Sci.* 24 (1976): 353-55.

20. William. R. D., and G. F. Warren, "Competition between purple nutsedge and vegetables," *Weed Sci.* 23 (1975): 317-23.

21. Knake, E. L., and F. W. Slife, "Giant foxtail seeded at various times in corn and soybeans," *Weeds* 13 (1965): 331-34.

22. Weatherspoon, D. M., and E. E. Schweizer, "Competition between kochia and sugar beets," *Weed Sci.* 17 (1969): 464-67.

23. Anonymous, "Metabolites of plant pathogens may prove useful in weed control," *Crops and Soils* 32 (4) (1980): 26.

24. Gajić, D., S. Malencić, M. Vrbaskì, and S. Vrbaskì, "Study of the possible quantitative and qualitative improvement of wheat yield through agrostemin as an allelopathic factor," *Fragm. Herb. Jugoslavica* 63 (1976): 121-41.

25. Zabyalyendzik, S. F., "Allelopathic interaction of buckwheat and its components through root excretions," *Vyestsi Akad. Navuk BSSR Syer Biyal Navuk* 5 (1973): 31-34 (in Belorussian, Russian summary).

26. Lykhvar, D. F., and N. S. Nazarova, "On importance of legume varieties in mixed cultures with maize," in *Physiological-Biochemical Basis of Plant Interactions in Phytocenoses,* ed. A. M. Grodzinsky (Kiev: Naukova Dumka, 1970), 1:83-88 (in Russian, English summary).

27. Eidt, D. C., and C. H. A. Little, "Insect control by artificially prolonging plant dormancy—a new approach," *Can. Entomol.* 100 (1968): 1278-79.

28. Scholze, P., "Untersuchungen zum Einflus trophischer Faktoren auf die Entwicklung der Blattläuse, speziell der Schwarzen Bohnenlaus, *Aphis fabae* Scop. (Homoptera, Aphidina)," *Zool. Jb. Syst.* 98 (1971): 455-510.

29. Scheurer, S., "The influence of phytohormones and growth regulating substances on insect development processes," in *The Host-Plant in Relation to Behaviour and Reproduction,* ed. T. Jermy, Symp. Biol. Hung. 16 (New York: Plenum Press, 1976), pp. 255-59.

30. Williams, C. M., "Third-generation pesticides," *Scient. Amer.* 217 (1) (1967): 13-17.

31. Horn, D. H. S., "The ecdysones," in *Naturally Occurring Insecticides,* ed. M. Jacobson and D. G. Crosby (New York: Marcel Dekker, 1971), pp. 333-459.

32. National Academy of Sciences, *Pest Control: An Assessment of Present and Alternative Technologies,* vol. 1, *Contemporary Pest Control Practices and Prospects* (Washington, D.C.: National Academy of Sciences, 1975).

33. Jones, R. L., W. J. Lewis, H. R. Gross, Jr., and D. A. Nordlund, "Use of kairomones to promote action by beneficial insect parasites," in *Pest Management with Insect Sex Attractants and Other Behavior-Controlling Chemicals,* ed. M. Beroza (Washington, D.C.: American Chemcal Society, 1976), pp. 119-34.

34. National Research Council, *Principles of Plant and Animal Pest Control,* vol. 5, *Vertebrate Pests: Problems and Control,* report of Subcommittee on Vertebrate Pests, National Academy of Sciences (Washington, D.C.: 1970).

35. Takahashi, M., "Physiological and ecological studies of *Lycoris radiata* Herb. 2. Prevention of mice-boring to field levees," *Weed Study* 25 (1980): 6-9 (Japanese, English summary).

Index

A

Abrus precatorius: see rosary
  pea
Acanthoscelides obtectus: see
  bean weevil
Aesculus octandra: see horse
  chestnut
Agropyron repens: see couch
  grass
Agrostemma githago: see corn
  cockle
Agrotis ipsilon: see cutworm,
  black
Ailanthus altissima: see tree of
  heaven
Alcohols: 46, 66, 72, 90, 123
Alders: 17, 47
Aleurites fordii: see tung tree
Alfalfa: 12, 13, 32, 117, 127
Alkaloids: 76-77, 84-85, 88, 90,
  96, 117-20, 127, 184
Allelochemics: definition, 22;

multiple roles of, 174-76
Allelopathic crop plants:
  decomposing residues, 23-32;
  crops that inhibit weed
  growth, 42
Allelopathic weeds: 37-41
Allelopathy: definition, 13;
  how toxins leave plants,
  33-34; biological weed
  control, 42-44; bridges
  between plants, 44-45; role
  in plant diseases, 45-49
Amino acids: 96, 117, 184
Amrasca devastans: see jassid
Ants: 16, 72
Aphids: 85, 90, 180
Apis mellifera: see honeybee
Apple: 12-15, 55, 90, 137
Argyrotaenia velutinana: see
  leaf roller, red-banded
Armyworms: 98
Arsmart, hot and mild: 16
Arum plant family: 108-109

33
Potato psyllid: 125
Prickly ash: 121
Protoparce sexta: see tobacco
hornworm
Prunus persica: see peach
Pteridium aquilinum: see
bracken fern

Q

Quassia: 20
Quercus spp.: *see* oaks
Quince: 137
Quinones: 78, 83, 95

R

Rabbits: 162-66
Radish: 4, 5, 33
Ranunculus spp.: *see* butter-
cups
Rape: 16, 71
Rats: 16, 166-68
Rattlebox: 53, 56-57, 178
Reed: 5
Reed canary grass: 155
Rethrins: 83, 119-20, 124, 126
Rhizobitoxine: 42
Rhododendron: 7
Rice: 29-31, 183
Robinia pseudoacacia: see
locust, black
Rockweed: 18
Rosary pea: 115
Rose of sharon: 85

Rotenoids: 114, 120
Rue: 5-6, 17
Rumex crispus: see dock, curly
Ruminants, effects of sage-
brush, orchard grass, and
sorghum on: 160-61
Rye: 24, 25, 27, 29, 42, 54, 55,
115
Rye, Italian: 7

S

Sage: 6
Sapindus spp.: *see* soapberry
Saponins: 77, 90-91, 96, 117,
184
Scabiosa: 7
Schinus molle: see piru
Schizaphis graminum: see
greenbug
Scrobipalpa ocellatella: see
beet moth
Secale cereale: see rye
Senecio spp.: *see* groundsel
Setaria faberii: see foxtail,
giant
Sheep: 152
Silkworms: 74, 81
Simmondsia chinensis: see
jojoba
Sinapis arvensis: see crunch-
weed
Sitophilus zeamais: see weevils,
rice
Slugs: 150-51
Snails: 19, 150-51
Soapberry: 116

67, 75

Wheat: 5, 7, 12, 19, 23, 25, 27, 32, 40, 42-43, 115, 177, 182

Wheatgrass *(Agropyron glaucum):* 43

Wolfsbane: 15

Woollybear, yellow: 98

Wormseed: 59; mustard, 42

Wormwood: 13, 14, 16

X

Xanthoxylum clava-herculis: see prickly ash

Y

Yew: 7

Z

Zea mays: see corn

Pest Control with Nature's Chemicals,

designed by Bill Cason, was set in versions of Times Roman and Helvetica by the University of Oklahoma Press and printed offset on 60-pound Glatfelter B-31, a permanized sheet, with presswork by Thomson-Shore, Inc., and binding by John H. Dekker & Sons.